WOLF ISLAND

ALSO BY L. DAVID MECH
PUBLISHED BY THE UNIVERSITY OF MINNESOTA PRESS

The Wolf: The Ecology and Behavior of an Endangered Species

The Wolves of Denali
(with Layne G. Adams, Thomas J. Meier,
John W. Burch, and Bruce W. Dale)

WOLF ISLAND

DISCOVERING THE SECRETS
OF A MYTHIC ANIMAL

L. David Mech

with Greg Breining

Foreword by Rolf O. Peterson

University of Minnesota Press
Minneapolis
London

Published by the University of Minnesota Press
111 Third Avenue South, Suite 290
Minneapolis, MN 55401-2520
http://www.upress.umn.edu

ISBN 978-1-5179-0825-6 (hc)
ISBN 978-1-5179-1131-7 (pb)

A Cataloging-in-Publication record for this book is available from the Library of Congress.

Printed in Canada on acid-free paper

The University of Minnesota is an equal-opportunity educator and employer.

25 24 23 22 21 20   10 9 8 7 6 5 4 3 2 1

For Donald E. Murray (1928–2014), a superb, skilled bush pilot who served the Isle Royale wolf and moose research project from 1959 to 1979. Don ensured that researchers safely gathered every last piece of information they could glean from the vantage of a light aircraft circling low over the island's bush.

CONTENTS

FOREWORD

Rolf O. Peterson

In February 1959, flying in a small plane over Isle Royale National Park, Dave Mech glimpsed primeval nature. Wolves were engaged in a struggle for survival that meant preying on moose, ten times their size. These species have been so engaged through evolutionary time, molding each other, yet we knew little about their relationship until Dave took to the air and witnessed it. Dave's initial observations of wolves hunting moose were nothing short of spectacular, and in 1963 millions of readers of *National Geographic* saw his photographs shot from aircraft and learned about Isle Royale for the first time.

Dave's research task, an all-consuming activity for three years, was to figure out the truth about the ability of wolves to kill prey. Would they whittle down the moose population to the point where its survival might be jeopardized? What factors might permit coexistence of predator and prey, both more ancient than the human species? In the superlative observatory that Isle Royale provides, Dave provided outstanding first-cut answers to many important questions. His research established field methods and a foundation that continues to frame research six decades later. Perhaps even more important, Dave was instrumental in creating a sea change in public attitudes about wolves, from evil vermin to respectable fellow travelers. Throughout his career, Dave has worked to discover and share the truth about this remarkable predator.

When I read this memoir of Dave Mech's first three years of wolf–moose fieldwork in Isle Royale National Park, I was reminded anew about the amazing descriptions of wolves hunting in open-forest habitats, the graphic accounts of moose cut down by wolf predation, and the grisly on-the-ground examinations of recently killed moose. For me, the overwhelming impression was, simply, this: trees grow, forests change. The young forests of Dave's time, considered prime habitat for moose, following wild-fires in 1930s and 1940s, were on the way out in the decades that followed Dave's pioneering study. Never again would researchers observe Isle Royale wolves and moose in the young, open forests that Dave witnessed. The same habitats supported sharp-tailed grouse as well, a prairie-edge species that was doomed to disappear as the trees grew.

Indeed, the first wild animal Dave saw on his initial flight in February 1959 was a sharp-tailed grouse, indicated on his penciled flight line on a map that now resides in the University Archives at Michigan Technological University. I marveled at this when I first saw it, as I in turn witnessed the last remaining dancing grounds for this bird when I followed in Dave's footsteps little more than a decade later, in the early 1970s. The openings that this bird requires had disappeared, and with that the bird itself died out.

That should have been the fate of moose as well, if the population theories of wildlife biologists of the day had been correct. In the mid-twentieth century the common notion in wildlife ecology was that populations of wild animals were a function of their habitat. The forest fires on Isle Royale in 1936 and 1948 provided optimal habitats for moose for a couple of decades, but then forest succession led to older forests less productive of moose forage, leading to the expectation, written out in my own PhD thesis in 1974, that moose would in turn decline. This is the essence of a "bottom-up" world.

This is not how things turned out. In 2019, with much older forests and a reduced habitat capacity, there are several times as many moose on Isle Royale as there were sixty years ago. How do we explain that? The answers would require decades of follow-up work, work that is still ongoing in 2020. We now understand, at the risk of oversimplification, that wolves, moose, and vegetation comprise three trophic levels, and each level influences and is influenced by the adjacent levels up and down this "simple" food chain.

This book highlights the important human element of the research, and here Dave details the privilege and huge responsibility that he undertook. After the first winter flight, Dave wondered if he could last through three winter seasons of being airsick. I wondered the same thing after my first flight in 1971, sitting behind the already legendary pilot Don Murray. Yet the first view of wolves from the back seat of Don's plane has remained one I have never forgotten.

Most appropriately, this book is dedicated to Don Murray, who served as the winter study pilot for nineteen years. His absolute dedication to discovery was a major reason Dave's initial three years of research led to something extraordinary—the longest study of a predator–prey system ever undertaken. After 1979, Don was unable to fly after suffering injuries in a car accident, and he was later to lose sight in both eyes. But he never forgot the amazing observations of wolves and moose from his little airplane, recounted with obvious pleasure every time I talked to him, even on the day he died in 2014. The last time Don talked to his grandson, also named Don and by then piloting the winter study at Isle Royale, he said, "All I care about are the wolves and moose." With that he passed the torch.

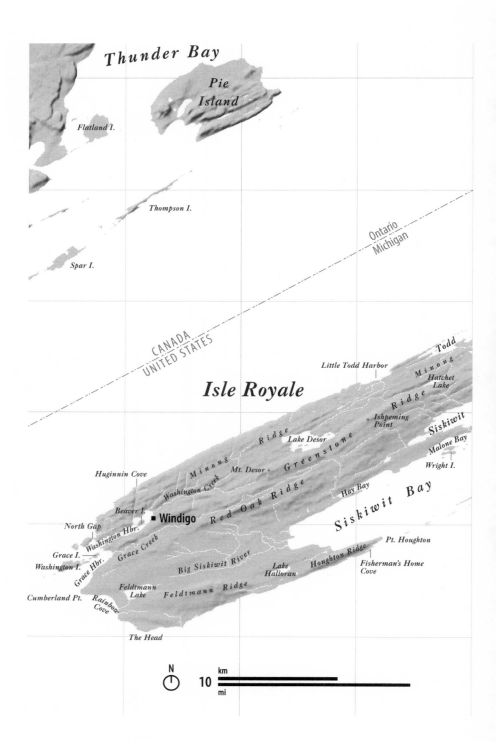

Thunder Bay

Pie
Island

Flatland I.

Thompson I.

Ontario
Michigan

CANADA
UNITED STATES

Isle Royale

Spar I.

Little Todd Harbor

Todd

Minong

Hatchet
Lake

Ridge

Siskiwit

Minong

Ridge

Lake Desor

Greenstone

Ishpeming
Point

Huginnin Cove

Washington Creek

Mt. Desor +

Malone Bay

Wright I.

Red Oak Ridge

Hay Bay

Siskiwit Bay

Beaver I.

Windigo

North Gap

Washington Hbr.

Grace I.

Washington I.

Grace Creek

Grace Hbr.

Pt. Houghton

Big Siskiwit River

Lake
Halloran

Houghton Ridge

Fisherman's Home
Cove

Cumberland Pt.

Rainbow
Cove

Feldtmann
Lake

Feldtmann Ridge

The Head

N
⊕
10

km

mi

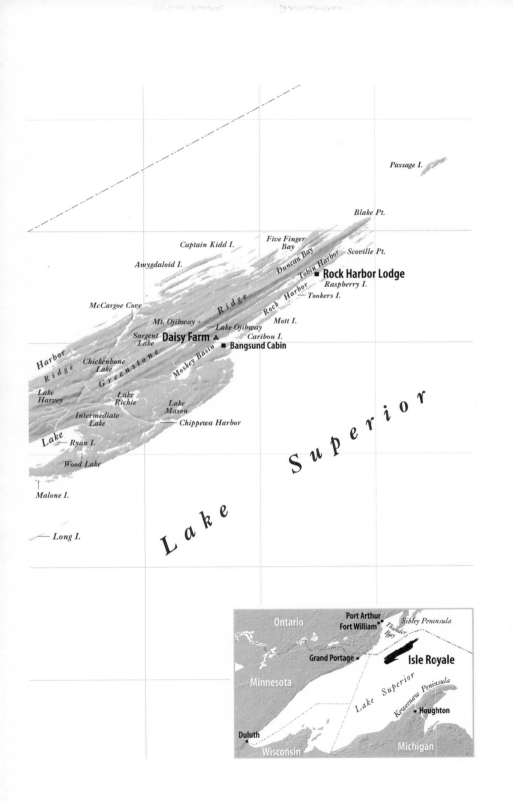

Passage I.

Blake Pt.

Captain Kidd I.

Five Finger
Bay

Duncan Bay

Scoville Pt.

Amygdaloid I.

Tobin Harbor

Rock Harbor Lodge

McCargoe Cove

R i d g e

Rock Harbor

Raspberry I.

Tookers I.

Mt. Ojibway +

Lake Ojibway

Mott I.

Sargent
Lake

Daisy Farm ▲

Caribou I.

■ **Bangsund Cabin**

Harbor

Chickenbone
Lake

G r e e n s t o n e

Moskey Basin

R i d g e

Lake
Harvey

Lake
Richie

Lake
Mason

Intermediate
Lake

Chippewa Harbor

L a k e

Ryan I.

S u p e r i o r

Wood Lake

Malone I.

L a k e

Long I.

Ontario

Port Arthur
Fort William

Sibley Peninsula

*Thunder
Bay*

Grand Portage

Isle Royale

Minnesota

Lake Superior

Keweenaw Peninsula

Houghton

Duluth

Wisconsin

Michigan

WE DIDN'T KNOW

꩜

Flying low beneath a sky of leaden stratus clouds, we searched the snowy forest and frozen shoreline of Isle Royale, the 45-mile-long island national park in northern Lake Superior. The leaves were down, of course—it was early February 1960—and the aspen and birch sprouted like thin gray hairs above the snow. But for scattered stands of spruce and balsam and pockets of dense cedar in small wetlands, there weren't many places a pack of sixteen wolves could hide.

We had spotted them yesterday on Lake Richie, one of many small interior lakes on the central part of the island. But today we had trouble finding them or picking up their tracks in the snow. Don Murray, my pilot, had even landed the small ski plane so I could follow the trail on foot. The tracks led north through dense cedar swamps. That explained why we couldn't follow the trail from the air. Once I determined the wolves had headed west along the Greenstone Ridge, Don landed, picked me up, and we resumed our search.

We found the pack on Siskiwit Lake, strung out in a long line as they walked on the ice, heading toward the south shore. From what I had been able to tell on the ground and what we had been able to see from the air, they hadn't killed a moose yesterday or last night. They were probably on the hunt for new prey.

Unfortunately, we were running low on gas. Don banked the Aeronca Champion and returned to Washington Harbor to re-fuel. In less than an hour, we were circling back over the wolves. They had climbed the ridge along the south shore of the lake into the woods, and when we finally spotted them, they were running toward a cow moose and two calves. A few wolves charged down off the ridge, as if to cut the moose off to the north. Two others ran toward the south.

The moose were racing along the steep ridge paralleling Wood Lake and dodged south through a burned-over area of aspen stubble toward a cedar swamp. The first two wolves were gaining rapidly and within a quarter mile from where the chase had begun had caught up to them, one wolf running on either side of the cow and her two calves.

The cow ran closely behind the slower calf. A couple of times she feinted toward the wolves, which, well aware of the power of her hooves, jumped out of the way. But each time, they quickly caught up. Most of the pack had closed the gap as well. The moose dove into a clump of cedar as four or five wolves tore at the rump and side of one calf and clung to it. Within 50 feet, the calf tumbled to the snow in a copse of cedars. The cow and other calf kept running, the wolves chasing them another 200 yards before pulling up and heading back toward the downed calf. The cow and remaining calf stopped and even drifted back toward their wounded relative.

Most of the wolves now bunched around the fallen calf, which hadn't moved from where it had dropped. From the air, we couldn't see through the cedars, but from what we could tell, within five minutes of falling, the calf was dead.

As gruesome—and admittedly, exciting—as this was, the kill was a rare opportunity. In nearly two years of fieldwork on the island, I had never witnessed a successful hunt. Nor had I ever had the opportunity to see a fresh kill up close.

"I want to get down there!" I shouted to Don over the noise of the wind and engine. He shot me a dubious look as he circled toward Wood Lake, less than a mile away.

I told Don I planned to hike in to the dead calf.

"Too dangerous," he said. Don, from Minnesota's Mesabi Iron Range, was a wolf hunter, one of the men of that era who hunted in planes and shot wolves from the air. Like many men who spent a lot of time in the woods, he believed wolves were dangerous. I wasn't so sure. From what I had read of wildlife biologists studying wolves—and admittedly, there weren't a lot!—I figured I could get away with it. Besides, I had a .38-caliber service revolver the National Park Service insisted I wear.

"Look," I said to Don. "Circle the plane over the kill. And if the wolves give me trouble, dive down on them and scare them away."

We glided to a stop on the ice. I clambered from the plane, strapped the revolver on my waist, and grabbed a canvas pack with camera, notebook, and a 16-mm spring-wound Bolex movie camera. I untied the snowshoes from the struts of the plane, stepped into the bindings, and shuffled across the lake toward the kill. The snow was only about a foot deep, but wet. Not the best for snowshoeing or walking. Don roared down the lake, the plane lifted into the sky, and he began to circle over the kill. To tell the truth, I was not sure it was a great idea either. But the thought of the plane overhead gave me some confidence.

I climbed the ridge on the lakeshore and headed into the woods toward the cedar swamp. Several wolves hung on the edge of the cedars, and when I got to within about 50 yards, they ran off. That was a relief. I continued snowshoeing toward the copse of cedars where the calf went down. Two wolves remained. But when I got to within about 50 feet, they sped off though the snow. So far, so good. I was going to get a look at the fresh carcass.

It lay in the snow, pretty much where it had fallen, about a half mile southwest of the southern tip of Wood Lake. Though clearly a calf, it weighed perhaps 300 pounds. The wolves had ripped its nose, neck, the left side of its chest, abdomen, and rump. The snow around the animal was red with blood. Even though it had been dead only about a half hour, the wolves had already eaten its nose and completely skinned its neck and the left side of its chest. The heart, part of the lungs, and the rump were already eaten.

I swung off the pack, took out a knife, and cut away the calf's mandible, a valuable piece of anatomy for later determining the health and age of the creature. I fetched out the cameras. I took movies and slides and black-and-whites from every angle I could think of.

3

Suddenly the roar of the plane grew louder. Don was diving directly at me! I looked around. Two wolves were bounding through the snow straight toward me and the kill. My mind swirled: the movie camera? Or the revolver? An opportunity? Or real danger?

I didn't know, of course. Nobody knew.

We knew that wolves survived largely by eating large hooved mammals. We knew they traveled in packs. We knew they once inhabited nearly all of North America but that in the last half century had been exterminated from the Lower 48, except for a few hundred in the wilderness of northern Minnesota, perhaps a scattered few in northern Wisconsin and the Upper Peninsula of Michigan, and a handful on Isle Royale in the midst of Lake Superior. The rest of the continent's wolves lived in Canada and Alaska.

Otherwise, we didn't know much about wolves. Even scientists. A few biologists had studied wolves in the preceding decades—Adolph Murie around Mount McKinley (now Denali) in Alaska, and Sigurd Olson and Milt Stenlund in northern Minnesota. But really, we were just getting started. We still didn't know much about the organization of wolf packs, how far they traveled, how they hunted, whether they could live on a variety of foods including small mammals, or to what extent they actually affected the populations of their chief prey species. It was just too hard to count them or to actually catch them in the act of hunting. They traveled too fast and far for a human on foot to stand a chance of keeping up. Even the wide availability of light aircraft had so far simply enabled biologists to spot them, note their location, and find their old kills. Radio collars and telemetry were still a pipe dream.

But I was here to add something new to the scientific knowledge of the wolf.

I was a new, enthusiastic graduate student at Purdue University. I had been recruited by my advisor, Durward Allen, a professor of wildlife ecology and natural resources and author of *Our Wildlife Legacy*. Allen realized that at a time when gray wolves had disappeared from all but a few isolated areas of the Lower 48, Isle

Royale's 210-square-mile ecosystem provided an unprecedented opportunity to learn more about this still mysterious symbol of wilderness.

Beginning in 1958, we plotted out a multiyear project to better understand how wolves hunted, what they ate, how their numbers tracked with the abundance of their primary prey, and how successfully they killed. (If one believed their critics, and even many biologists, wolf packs were killing machines that could catch their prey at will.)

This is the story of the first three years of that study—the time I spent on the island, hiking hundreds of miles on trails in summer, flying hundreds of hours over the island in winter, learning all I could about the habits of wolves and the island's only significant wolf prey, the majestic moose.

I had the opportunity as well of seeing a vanishing way of life—getting to know the last few families of commercial fishermen living on the island, tracking perhaps one of the last lynx to be found on Isle Royale, and flushing sharp-tailed grouse, a bird more typically associated with brush and grasslands, that no longer lives on the island.

By the time I left, we had set in motion a research program that continues today, more than sixty years later. I was followed by a parade of researchers: Phil Shelton studied beavers. Pete Jordan and Michael Wolfe learned about moose population dynamics. Rolf Peterson picked up the project and guided it for more than forty years, a time when the island's wolf population underwent tremendous fluctuations. Peterson's former student, John Vucetich, continues to lead the project today.

The Isle Royale study has been the longest continuous predator–prey study in all of science. And it has been one of the most successful, adding immensely to our knowledge of wolves and moose and their habits and interactions. More profoundly, the Isle Royale fieldwork has upended our understanding of the "balance of nature" and cast serious doubts on our ability to predict the contingencies that drive wildlife population dynamics. More recently the island's wolf population was dying out, beset by inbreeding and lack of new recruits from the mainland. To reverse that fateful trajectory, the National Park Service in 2019 introduced new wolves from the mainland. The decision to interfere in the "natural" order of a national park was controversial, and success is far from assured, but the reintroduction has created a host of new opportunities to study wolves and moose.

But as I say, at the time we didn't know much. We didn't even know for sure if a man trudging through foot-deep snow could scare a pack of sixteen wolves off a freshly killed moose and expect to live to tell the tale.

So when those two wolves bounded toward the calf they had just killed and this twenty-three-year-old, budding but naive biologist standing over it, with a movie camera in one hand and the .38 in the other, I didn't know whether to raise the movie camera or the revolver.

I raised the gun. And the moment I did, the two wolves, just 50 feet away, skidded to a stop, whirled around, and sprang off running. I still regret my choice. I put the pistol away for good and quit carrying it altogether.

Perhaps I shouldn't have generalized from that single moment, but I grew confident not only that my life would be safe from wolves but that this project was destined to produce a wealth of valuable science for the future.

THE CHANCE OF A LIFETIME

⁂

W hen I first heard of the island, it sounded most mysterious.
Could there really be a huge chunk of rocky wilderness
lying far out in Lake Superior?

I was twenty years old and just beginning my senior year at
Cornell University. With the exception of a few short trips, I had
spent my entire life in upper New York State. I realized there were
many parts of the country I knew little about. Maybe there *was* a
large wilderness island, a world of its own, right within this heavily settled country of ours.

Another thing that helped fire my imagination was the name of
this isolated wilderness: Isle Royale, the royal island. It sounded
majestic. Isle Royale was a national park, but from what I could
gather, few people ventured there. It had no roads, so visitors arrived by boat and traveled on foot on its trails, or by water along
the shoreline. No one lived there for most of the year. But the
most intriguing fact I learned—and the detail most relevant to my
work—was that Isle Royale was inhabited by moose and timber
wolves.

The guy giving me my first fuzzy image of Isle Royale was
a graduate student who had just come to Cornell from Purdue
University. He told me Durward Allen was planning a study of
the wolves and moose of Isle Royale—if he could get the money.
What a great project that would be, I thought.

I probably had a keener feeling than most people about just how
fascinating a research project on wolves could be because of my
own interests. My major field of study at Cornell dealt with wild
animals, how they behaved and interacted with their surround-

7

ings. I was most interested in mammals—deer, rabbits, foxes, mice, woodchucks, beavers, and the like. And of these, it was the carnivores, the animals that killed their own food, that interested me most.

I could get enthused about working with these animals wherever they were found, but I most liked studying them in their native, natural surroundings. An ideal job, to my mind, would involve the study of wilderness carnivores. And so, while I was interested in weasels and mink, their wilderness relative the fisher fascinated me even more. Among the dog family, I was fond of working with foxes, but I much preferred snowshoeing through the Adirondack Mountains in search of their larger, wilder cousin, the coyote, or brush wolf.

And the wild dog that interested me most of all was even larger than the coyote and lived only in the wildest regions of the North—the timber wolf.

But news of a wolf project to be carried out by an eminent biologist on a far-off island was of no great consequence to a lowly undergraduate with no particular talent, barely able to maintain respectable grades. Oh, perhaps someday I could sit in awe and read about the study, just as we students pored over Adolph Murie's classic *The Wolves of Mount McKinley*.

During the next several weeks I became fully preoccupied with my own problems and responsibilities. Life as a student hadn't been easy. Twice during my stay at Cornell I had come close to dropping out because of the pressures of studying, and my senior year was turning out to be especially difficult.

At the time, there were the usual six courses per semester to attend and prepare for, which was more than a full-time job. Then there were the thirty hours of work per week in a local grocery store. The duties of the presidency of the Cornell Conservation Club took additional time. But on top of all that, as a senior I had to plan and prepare to gain admission to a graduate school for the next year.

Choosing a graduate school, and getting accepted, is a crucial step in a student's life, for it can easily determine his or her future for many years. One must try to find a good school, a sharp and experienced major professor to work with, financial help in the form of an assistantship or fellowship (a part-time job teaching or doing research), and an interesting study to carry out and write a thesis about. That's a pretty tough combination to find when you are competing with umpteen other students from the rest of the colleges and universities in the country, all looking for the same thing.

I really had little idea where I would like to go or what I would like to do—except that whatever my choice, it would have to lead to a career in ecological or behavioral research on animals in their natural surroundings. This much I had learned during the summers of 1956 and 1957 and a few weeks in 1958 as a field assistant on a black bear study run by Cornell University and the New York State Conservation Department.

The bear project had been an excellent experience. We live-trapped black bears both at dumps and in wilderness areas in the Adirondack Mountains. Once captured, the bears were knocked out, ear-tagged, weighed, measured, examined, and released. Then, if we trapped them again, or if they were shot or killed by a car, we would learn how far they traveled and could estimate how large their territories might be. Imagine—all that effort hoping to obtain two locations for a few bears. However, at that time, even two locations for a single animal produced new information.

We captured the garbage-dump bears in "culvert traps"—a section of culvert mounted on a trailer. We anesthetized the bears by spraying ether into the barrels. But the bears in the wild were another story. We caught them in steel foot-hold traps. To subdue them, one of us slid a chain loop, or "choker," around its neck and snugged it tight. As one man held the bear by the choker, another looped rope around each of the animal's legs. We tied the ropes out to trees until the bear was spread-eagled. Then one of us would inject the bear in the abdomen with sodium pentobarbital. Within ten minutes, the bruin was sound asleep. I got pretty good at it, even as an undergraduate. (Nowadays, with new types of drugs, bears can be easily anesthetized with a dart gun or a syringe on the

9

end of a "jab stick," injecting the drug into muscle—easier and less traumatic for bear and researcher.) Once we had the bear drugged, we could ear-tag it, weigh it, measure it, sometimes paint it for visual recognition, and even remove one testicle for histological studies. The local lumberjacks thought we were trying to cut the bear population in half!

During the summer of 1956, between my sophomore and junior years in college, I was part of a four-man crew that trapped at dumps almost exclusively. But the next year, I was put in charge of a wilderness crew—a crew of just two of us. We trapped old logging roads in the central Adirondacks from dawn till dark, often eating dinner at midnight. We kept that pace through the entire summer without a day off. The two of us trapped and released nearly fifty bears.

Doing this work, I learned a great deal not only about bears but also the wilderness and animals such as fishers, bobcats, and coyotes, all of which we sometimes caught in our traps. But most of all, I learned that this was the kind of work I was happiest with. It combined the beauty of working outdoors with the tremendous satisfaction of learning about splendidly interesting creatures. This had to be my life's work.

That's why it was so important for me to make the right arrangements for a graduate education. Researchers who want to work on their own ideas should have a PhD, and that takes at least four years in graduate school.

But two months into my senior year I still hadn't settled on a graduate school. There *was* the possibility of studying snowshoe hares in Canada. And a kind of uncertain fox project somewhere else. And there was a good chance of joining a coyote study for my graduate work.

About the middle of the fall semester, one of our classes traveled to a statewide meeting of several conservation groups. The main speaker was to be Dr. Durward Allen of Purdue. I had heard Allen speak once before and had admired him ever since reading his entertaining and informative book *Our Wildlife Legacy*. I looked forward to hearing him again.

At the meeting, our class sat in on several speakers and discussions. I became especially engrossed in a talk on the preservation of the Adirondack wilderness, a subject of great concern to me. The Adirondacks made up the one large wilderness left in the East, and I spent as much time there as possible. Any threat to the wild character of the area disturbed me.

Suddenly someone tapped my shoulder. Ollie Hewitt, my professor at Cornell, motioned for me to get up and follow him.

"C'mon, I want you to meet Dr. Allen," he whispered as we sneaked out of the room.

I was flabbergasted. Why would Dr. Hewitt pick me, the only guy in class who hadn't thought to wear a white shirt and tie to the conference, to meet Durward Allen? When it came to meeting famous people, I felt as awkward as a teenager.

And so it was one bewildered, shabbily dressed, and ill-at-ease kid who shuffled into a large room where sat Durward Allen. He invited me to sit down. He was not a large person, and he had a friendly voice. He punctuated his words with a smile, and I began to relax.

"What are your plans for graduate school, Dave?"

I told him I definitely wanted to go and that I hoped to find some sort of project on carnivores.

"Dr. Hewitt tells me you have worked a couple of summers on the bear project in the Adirondacks."

"Yes, and I really enjoyed it," I said. "That's the kind of work I would like to do when I get out of school. It's a great pleasure to work in the wilderness and to be helping to uncover brand-new information about animals."

"What about winter?" he asked. "Have you ever done much work in the wild during winter?"

I was glad he asked. I told him that for me winter was the most exhilarating time of year, especially in the wilderness. Several college buddies and I had spent nearly every Christmas and Easter vacation pursuing furbearers in the Adirondacks. Over Christmas, we would don snowshoes and track fishers over the mountains days on end, often traveling many miles a day. When one of these

10-pound weasels finally holed up in a hollow tree, we would chop it out and capture it for its fur. During the Easter break, we spent many futile days trying to trap otters and beavers in those same mountains.

Dr. Allen told me about a new project he was hoping to begin soon. It would require a lot of winter work to study wolves and moose on a large wilderness island in Lake Superior.

I realized I was about to hear the details of the new wolf project that older grad student had mentioned several weeks earlier. But now the story was coming directly from the guy in charge.

Moose had lived on Isle Royale since they swam there in the early 1900s, decades before wolves found their way to the island, Allen said. Without a predator capable of killing them, moose multiplied and began eating themselves out of house and home. In the mid-1930s, a massive die-off took place as moose went hungry. By the late 1940s, the herd had built up again and had begun starving once more. Wolves appeared on Isle Royale about 1949, probably by crossing the frozen lake in winter.

Since the arrival of wolves, not much was known about their numbers, their behavior, or their effect on the moose population. Clearly, several wolves continued to inhabit the island, and they were killing moose. Some people even feared they would kill all the moose. It was also known, from a short study conducted there in the mid-1950s, that both moose and wolves could be counted, tracked, and observed from a light aircraft in winter when the trees and underbrush were bare.

Dr. Allen hoped that a biologist flying in a light plane would be able to discover new details about the wolves and moose. How many wolves are on the island? How many moose? How many moose are the wolves killing? How fast is the moose herd reproducing? How fast are wolves adding to their numbers? Will wolves kill off the moose? Or will the populations of moose and wolves achieve some sort of equilibrium? These were only a few of the questions he thought might be answered.

It was a study he had wanted to do for several years, Dr. Allen said, but only recently did it look like he might get funding. He imagined the study lasting at least ten years. Unfortunately, because of his many commitments, he wouldn't be able to carry out all the research himself. He planned to make a series of graduate student projects the backbone of the work.

I think I was almost as happy as Dr. Allen was when he told me that his chances to begin the study seemed very good. It sounded even more fascinating to me now that I had heard the details. But I did wonder what any of this had to do with me. That's why his next sentence floored me.

"I would like you to do the first study."

I could not have been more surprised. Despite the interview of leading questions, it had never dawned on me that he would consider me for such a job. It didn't make sense. Why would Durward Allen want *me* to carry out this study?

I struggled to recover. I asked various questions about the project, about Purdue, about courses, about anything that would help cover up my shock and surprise.

But it turned out this was for real. Dr. Allen even advised me to double up in French for the coming spring semester so that I would have a start on the language requirement for graduate school. I was also supposed to take various exams through which I might win a fellowship to support me through part of the wolf work.

We talked for perhaps an hour and a half. I was still reeling when we were done. Dr. Hewitt and I then returned to the conservation meeting while Dr. Allen gave his speech. I have never been able to recall what he talked about.

During the following months I corresponded several times with Dr. Allen and followed his advice. For a while it wasn't certain he would get money for the project, but sometime in the spring he received word that he had been given the green light by the National Science Foundation. In addition, I had been accepted by the Purdue Graduate School and had been granted a fellowship.

That meant all I had to do was get through my spring semester, including my double-up French course, and I would be able to begin the wolf project the following summer.

It turned out that getting through the last semester was no easy task. For one thing, I had fallen in love with a Cornell student, and we became engaged to be married after I returned from my first summer on the island. I didn't know how a wife or family might fit into the project, but we decided to figure that out after I explored the island and consulted with Dr. Allen. Would a marriage or family weather the weeks and months I would spend doing research?

More serious for the project was a strong recommendation from the college physician that I quit my job at the grocery store. I had injured a disc in my back pulling a 300-pound bear out of a culvert trap the previous summer. The disc was causing a change in the muscles of my left leg, and the whole situation could get worse if I continued to lift heavy cases of cans in the grocery store.

I took the doc's advice and quit my job, but that meant coming up with another source of money. So I sold my beautiful 12-gauge pump shotgun for $65. I wrote an outdoor story and sold it to *Sports Afield* magazine for $75. Then I got a $250 grant from the college, for which I was most grateful. By managing this money carefully and depending on fish (smelt from the local Cayuga Lake run) and road-killed rabbits and deer I found around the area, I managed to survive the last semester.

That took care of every problem that got in the way of doing the Isle Royale study except one.

The college physician had also recommended that I not attempt the wolf project. He thought I might further damage my back and rupture the disc, and that I then would have to have it removed. This kind of ailment can be crippling, he said. In fact, he said the best advice he could give was to urge me to seek a desk job.

He was asking me to give up the chance of a lifetime.

The decision wasn't easy. But despite the risk to my health, I felt I couldn't afford to pass up this opportunity.

So in June 1958, after breaking in some new workers on the bear project, I prepared to head for the royal island. I took a bus to West Lafayette, Indiana, and met Dr. Allen at Purdue. It was somewhat distressing to find that the university was situated in the middle of the Indiana cornfields. But it was consoling to know that the schedule for the next three years called for about half my time to be spent on Isle Royale.

Dr. Allen and I packed our gear in a university car and started north. In Wisconsin, we picked up Dr. Douglas Pimlott, who was about to begin a wolf study in Ontario. He was traveling with us to Isle Royale to join a group of other biologists knowledgeable about wolves and moose. These men were to form an advisory committee that would travel with us to the island for a few days, help give me more background, and perhaps suggest some lines of approach to the study.

The next morning we boarded the *Ranger II* in Houghton, Michigan. The boat, a 114-foot wooden-hulled surplus mine-layer, had made hundreds of passages from the National Park Service headquarters, through the Keweenaw Waterway to Lake Superior, and 60 miles over the lake to Isle Royale.

On the boat I met the rest of the advisory committee. Larry Krefting was a U.S. Fish and Wildlife Service biologist who had done several early studies of moose and moose food on Isle Royale. Robert Linn was a National Park Service naturalist and a student of the island's vegetation. Milt Stenlund worked for the Minnesota Department of Conservation and had authored a study of wolves in northern Minnesota. Ray Schofield studied predators for the Michigan Department of Conservation. Rodger "Rod" Stanfield was director of wildlife research in Ontario. C. Gordon Fredine and Bob Rose both worked for the National Park Service. I was thrilled that such accomplished people were meeting there to give me a good send-off.

For the first few hours on the boat we spent most of our time talking. But after awhile, we began watching the horizon for Isle Royale to materialize out of the haze.

"There it is!" someone exclaimed.

And there it was—a long, low bulge on the gray horizon, perhaps 20 miles away.

I was impressed by the size of this far-off world. And as we drew closer, it grew longer and longer. The island is only 9 miles

15

across at its widest point but stretches 45 miles long, running roughly southwest to northeast. We were approaching it more or less broadside, and eventually it began to extend in each direction until it looked like another shore. We could no longer see the ends of it. It was so large it hardly seemed like an island at all. Our slow approach afforded an excellent introduction, for it allowed the awe I first felt upon hearing about the island to grow and grow, like the island itself.

As we drew within a few miles, more and more details took shape: Dense stands of spruce and fir interspersed with birch and aspen. A long, ragged backbone against the gray sky. A dark rocky shore. It was easy to imagine moose barging around through the brush, and wolves loping over the rocks even though I had never seen either in my whole life.

The subtle beauty of Isle Royale didn't begin to show until we were about a half mile away. Jagged reefs, rocks, and tiny islands—more than four hundred, I later learned—lay just off the shore of the main island. Most were colored a deep chestnut, but those poking up through the water were decorated with gaudy splashes of bright orange and light grayish-green lichens. Huge rolling breakers of clear water smashed the shoreline. This was indeed an isolated wilderness world.

My base for the summer would be Mott Island, about a quarter mile off the shore of the northeastern part of the main island, where the National Park Service had its summer headquarters, including a bunkhouse where I would stay when not on the trail.

Dr. Allen, some of the other wildlife experts, and I made several short trips around a small area of Isle Royale across the harbor from the headquarters. On one of the trails, Doug Pimlott, Ray Schofield, and I spotted a large dog-like track in the mud, and my heart throbbed at the proof that I was finally in timber wolf country. Not knowing just what might turn out to be important, I jotted down a quick note of its location.

Suddenly I became impatient. I wanted to ditch all my high-powered advisors and get off into the bush myself, into the interior

16

of this new world. I was eager to begin picking up the thousands of pieces of the wolf–moose puzzle and start putting them together.

That same day, Dr. Allen, the advisory committee, and I took a boat trip some 25 miles southwest along the shore to Siskiwit Bay. There the land was lower and flatter, and the birches, aspen, and willows were young and thick, the result, I learned, of an intense fire in 1936. This, I was told, was excellent moose country.

After we had settled into the old Civilian Conservation Corps camp where we planned to spend the night, a few of us ventured off to look around.

Just 200 yards from camp, in the waning hour of daylight, we spotted a bull moose. The huge dark-brown beast with massive velvety antlers stood tall in an open meadow and peered at us. It was my first moose, and I was impressed. The critter wasn't pretty. In fact, he was downright homely. Just standing and staring, he seemed even a bit stupid.

Finally, he lumbered off. It wasn't until a year and a half later that I came to understand the full meaning of his odd way of just standing stiffly rather than fleeing.

The next day we left Siskiwit Bay, and the various members of the advisory committee caught the boat back to the mainland. Just before he left, Milt Stenlund handed me a copy of his bulletin on wolves in northern Minnesota. On the first inside page, he had written, "To Dave: Good Luck on Isle Royale. Milt." Here was a real, experienced wolf biologist, the only such member of my advisory committee, wishing me good luck and tacitly showing faith in me, a young, green student. I felt really privileged.

Dr. Allen and I then planned out a four-day hike through the interior of Isle Royale. To begin, we boarded an old Park Service cabin cruiser used by the park's trail maintenance crew, which took us to Crow Point, about halfway along the southeastern shore of the island. We hiked up the Ishpeming Point Trail inland toward the Greenstone Ridge, the backbone of the island. Along this 6-mile stretch of trail, we were delighted to find fresh wolf tracks and several fresh and old wolf "scats."

The tracks were exciting to see. I had never seen them until this trip, and they provided a solid reminder of why I was here: Here's my subject! More accurately, here *was* my subject. A few tracks on a trail really said nothing more than that wolves had been here. And that much we already knew.

Scats, on the other hand, told a better story. Not only did they indicate the presence of wolves just as surely as a track, but they also showed what a wolf had been eating. By identifying the hair, bones, and other matter found in scats, I would be able to get a good picture of the food habits of the animal I was studying.

The wolf scats we found along the trail were packed with long, stiff, coarse hairs wrapped around chunks of bone. No doubt these were remains of moose. We dropped each sample into its own paper bag and labeled it with the date, location, and whether the scat appeared fresh or old. This was important because the fresh scats would be valuable clues to what wolves were feeding on during the season they were found. The old scats might have been months old, from any season, and were less useful in that regard.

To gain an accurate idea about the relative importance of various animals in the wolves' diet, I would have to collect hundreds of scats from all over the island and analyze them in a laboratory. Gathering scats, then, would be one of my main objectives during each of the three summers.

Dr. Allen and I reached the Greenstone Ridge, at a place called Ishpeming Point, about midafternoon. We stood more or less in the center of the 1936 burn. The surrounding brush was thick and lush but not very high. As we ate lunch on the ridge, we could look out to the south some 10 miles over Siskiwit Swamp, Siskiwit Bay, and a long highland known as Feldtmann Ridge.

Eight to 10 miles to the southeast and east we could see Malone Bay and several islands that form an archipelago outlining the far side of Siskiwit Bay. Beyond this lay only Lake Superior. The southern shore of Wisconsin and Michigan lay far beyond the horizon. There was not a sign of the hectic civilization we had left behind less than a week before.

Finishing lunch, we hiked the Greenstone Ridge Trail northeastward another 6 miles and found more fresh wolf tracks and scats. One scat was full of long, silky brown hair, small bones, and fine grayish fur. Beaver, we agreed.

We reached a rustic, Park Service lean-to on the shore of Hatchet Lake in the center of the island at six fifteen. We set up camp, cooked supper, and watched a pair of loons running across the surface of the placid water, flapping their wings, and uttering their strange and wild cries. We were the only humans within many miles, and it seemed as if they were putting on a special display just for us.

During the next two days, Dr. Allen and I continued northeastward along the Greenstone Ridge Trail, covering some 17 miles. We ended up at Daisy Farm, a clearing with several lean-tos along the southeastern shore of Isle Royale, where a small copper mine had operated more than a century earlier. Later, a sawmill cut lumber there, and a market garden grew vegetables for the Rock Harbor Lodge on the northeastern end of the island. These days it was a National Park Service campground. We poked around the site until a Park Service boat picked us up and brought us down the long finger of Rock Harbor back to Mott Island.

After spending a day making arrangements with park personnel for the first winter phase of the study, Dr. Allen took the *Ranger II* back to the mainland and returned to Purdue.

Finally, I was alone. Oh, there were Park Service employees and a few tourists visiting the lodge and hiking the trails, but in terms of the wolf project and my responsibilities for it, I really was on my own. This was a new experience. Even at college, I had been with friends, and home was only an hour away. But here I was alone with the wolves and moose. And the responsibility to succeed was squarely on my shoulders. I was in charge of a real scientific investigation in a place the likes of which I had never seen. It was a formidable yet wonderful prospect.

DISCOVERY

SUMMER 1958

꙰

So what was I to do this summer?
   Just knowing that I was in wolf range and seeing signs of the animals were a thrill. At one of the first tracks I found, I eagerly knelt down and measured: 3⅛ inches across and 4 inches from tip of middle nail mark to rear edge of the pad. Of course I longed to spot a wolf, but Dr. Allen had already warned me I was unlikely to see one until winter, when the dense foliage had died back. Still, around every bend and over every hill lurked the possibility of actually seeing the animal I was here to study.

   I had plenty to do, though none of it was as exciting as actually watching a wolf. On the long drive north from Purdue to our departure point at Houghton, the copper-mining town on Michigan's Upper Peninsula, Dr. Allen had told me that the most important thing to do this summer was to learn the island. Hike the trails, cruise the shore by boat, learn the topography, the plants, the animals. Along the way, there would be basic fieldwork: Take note of where you see moose, where you find fresh wolf tracks. Collect wolf scat to learn about their diet, and moose bones, especially jawbones, to learn about their age and state of health when they died.

   Dr. Allen, of course, was the person most responsible for directing my activities. He was a short, trim man. Personable and friendly, he was also given to somewhat cynical aphorisms. "The automobile is a great leveler," he would say. "Any fool can drive." I admired his writing. He believed it was important as a scientist to write for a lay audience. He enjoyed writing for popular magazines. In fact, that was an interest he and I shared, and we talked about it often. I had enjoyed writing since I was a kid, sending letters to the editor of the local newspaper. I took courses in writing for a popular audience, first at Cornell and then at Purdue. I liked the excitement of getting correspondence from a publisher, especially ripping open an envelope to find, not a rejection letter, but a letter of acceptance or even a check. I had recently written a story for *Pennsylvania Game News*—"Nature's Winter Layaway Plan"—about hibernation. Dr. Allen loved that title, and his enthusiasm was a source of pride and affirmation for me. Even better, I realized that as a student and now a graduate student, I was doing something interesting enough to write about. I had written about trapping fishers and bears in the Adirondacks for several publications. Now, studying wolves on a wilderness island in the world's largest lake was further grist for the mill.

To my way of thinking, Dr. Allen was the perfect advisor. Though he had never studied wolves himself, he was an expert in other aspects of wildlife biology. He had handed me a perfect opportunity and was willing to let me do almost whatever I wanted to do. His only directives: Count wolves and count moose. Find out what you can.

I spent the summer of 1958 acquainting myself with Isle Royale, some of it by boat along the shoreline, and much of it by hiking the network of trails in its interior.

Isle Royale is a ragged spit of rock, nearly ten times the size of Manhattan. Most of the island is made up of volcanic rocks, such as very dark basalt, with some sandstone-based conglomerates mixed in. The island is one end of a U-shaped geological formation that dips under Lake Superior and reemerges as the Keweenaw Peninsula of Michigan, which lies parallel to Isle Royale and, like the island, contains copper-bearing rock mined by both ancient Indians and more recent white settlers. Scattered around the main island are many outcrops forming hundreds of smaller islands and reefs, a considerable hazard to navigation and the source of many shipwrecks.

The geology forms a series of ridges running parallel to the length of the island, most evident in the Greenstone Ridge, which rises nearly 800 feet above the surface of Lake Superior and runs nearly the length of Isle Royale, and the long "fingers" that reach into the lake, forming several protected harbors at the northeast end of the island. The northwest sides of these ridges tend to form short cliffs and escarpments, while the southeast sides slope gradually, like dunes of volcanic rock. Indeed, the northwest shore of the island itself is hard and unforgiving, with few places of refuge. The southeast shore holds more bays and beaches.

Several small lakes sit in the rocky basins of Isle Royale. The largest is Siskiwit Lake, easily reached by a hiking trail on the south shore of the island. Others include Lake Desor, Hatchet Lake, Lake Harvey, Chickenbone Lake (shaped vaguely like a wishbone), Sargent Lake, Angleworm Lake, Lake Richie, and Intermediate Lake. I immediately liked the idea of a lake within an island within a much larger lake. And some of the lakes have their own small islands—like nesting dolls of lakes and islands.

Designated a national park in 1940, Isle Royale hosted a smattering of summer residents with lifetime leases, mostly commercial fishermen struggling to make a living since sea lampreys had invaded Lake Superior just twenty years before I arrived and had begun destroying the lake trout, lake whitefish, and ciscoes. Fishing may already have reached unsustainable levels, and the lamprey served to push the fishery over the edge. At Rock Harbor, at the northeast end of the island, the *Ranger II* arrived and departed. Rock Harbor was the main social center of the island, housing a ranger station and the staff employees of Rock Harbor Lodge. There was also a small recreation room where every weekend a Catholic priest came from the mainland and held mass. One could also rent outboard motorboats there to cruise around the harbors. Large yachts, too, would harbor there. At the southwest end of the island was Windigo, where there was also a small lodge and ranger station.

Isle Royale has no roads, but more than 100 miles of foot trails lace the island. The Greenstone Ridge Trail runs the rocky spine up the center of the island. Minong Ridge Trail follows the rocky heights along the north shore. Shorter trails climb from the shore to high outlooks and interior lakes and join the longer trails to form a network that covers the island. From my quarters at the park headquarters on Mott Island, I could cross the narrow channel to the main island, moor my boat, and spend several days on the trail either doubling back to the boat or heading to a destination where I could catch a boat ride from a Park Service ranger or trail crew back to my boat.

During my first summer on the island, I was single and allowed to stay in a bunkhouse with several of the Park Service workers, such as the trail crew, and various engineers and maintenance men. Although most of them disliked wolves and were convinced they were dangerous and would wipe out the moose, I got along with them fine. I often met the four trail-crew members out cleaning trails and sometimes stayed with them in a couple of patrol cabins in the island's hinterlands. Those were younger guys who were impressed with the miles of trails I hiked collecting scats and looking for wolf tracks.

As soon as Dr. Allen left, I hiked to Chickenbone Lake one evening after supper and walked along a nearby ridge overlooking the northern slope of the island, in hopes of seeing moose or wolves. After forty-five minutes I had seen none and hiked to a nearby, screened-in lean-to. It rained through the night, the roof leaked, and in the morning I built a fire to dry out my sleeping bag and clothes. Hiking to Hatchet Lake, I missed the trail junction and decided to head back to Chickenbone for the night. The next day I hiked to Ishpeming Point and ate and slept at the more primitive, screenless Lake Desor lean-to.

These hikes were a tremendous adventure. I loved the thump of my leather boots on the rocky, twisting trails. From the shore, I climbed the trails into the heart of the park, up Mount Ojibway. Emerging from the dense forest to a rocky prominence, I could climb a fire tower and see the blue sweep of Lake Superior all around. Carrying a canvas pack, a sleeping bag, cook kit, and some dried or canned food, I would spend the nights in the camping lean-tos. I would stash food at these locations so that in the future I could travel even lighter. When I wanted, I would hike all the way to Washington Harbor, more than 30 miles as the crow flies from headquarters. I was young, and as long as my back was healthy and strong and my disk wasn't acting up, I could manage 20 miles in a day with a 30-pound pack. During my first summer I learned to make my way to all these places on a schedule that allowed me to carry enough dried food to sustain myself for a few days at a time.

I took photographs, to illustrate my thesis and to help sell the articles I planned to write. I ate the blueberries and raspberries that grew on the exposed ridges. I scooped fresh water from springs and drank water out of puddles when I had to. To travel alone in the wilderness, far from the tourists who clustered around park headquarters, was exhilarating.

I was living out a childhood fantasy, a dream of traveling in the wilderness. When I was young, I had devoured *Robinson Crusoe*. Later were Lew Dietz's young-adult novels set in the woods of Maine—*Jeff White, Young Woodsman*; *Jeff White, Young Trapper*. I loved those books. My dad was a hunter and fisherman and would take my mother, younger brother, little sister, and me car camping in the Adirondacks. In winter, I would walk the few blocks from our house in Syracuse to climb Hunts Hill, all of 100 feet high, and look out over Onondaga Lake, imagining I was an explorer. In Boy Scouts I led my patrol in weekend hikes and cookouts. And of course, in college, I spent days with my friends, camping, snowshoeing, and trapping in the Adirondacks.

Isle Royale's terrain and forest were familiar from my days in upstate New York. I don't recall ever having been scared or apprehensive, even though I knew wolves and moose to be close at hand. I also knew that the naturalist Adolph Murie had once climbed into a wolf den, the mother wolf jumping out as he approached. How dangerous could it be?

As I walked and became accustomed to putting on 10 or 20 miles a day, spending the night wherever I was when the day ended, I began to think, yes, this is how wolves travel. It's easy. You just keep going. I knew wolves traveled a lot. Even humans can do it, just not quite as fast. It felt good to be traveling like a wolf. And that's what I did, day after day after day.

But the trails would take me only so far so fast. To more efficiently cover the area I was studying, I needed a boat, and I needed to learn how to pilot it. That was more daunting than it sounds. Superior may be called a lake, but it is no mere lake. Isle Royale is exposed to an unimpeded fetch of more than 100 miles from the southwest and southeast. Storm-driven waves often tower 10 feet (and sometimes much higher). Furthermore, the main island is surrounded by rocks and reefs that can tear the prop off a boat or open a gash in the hull. Many experienced commercial fishermen who worked the lake set out at dawn never to return.

Bob Linn, the park naturalist, came to the rescue, lending his 16-foot outboard runabout, dubbed the *Wolf*, to the project.

Bob was an interesting guy. Just over thirty, he cut a neat, soft-spoken figure with his flat-brimmed wool campaign hat and short, close-fitting Eisenhower jacket. In addition to being a PhD ecologist, he was also an artist. He lived on a small island near park headquarters and often invited people over for parties. I can say with confidence that Bob was the only person I met on Isle Royale who routinely and proficiently played the marimba. He had been coming to the park for years and just recently had been picked to be the chief naturalist. So far as I could tell, he wished never to leave the island.

Bob knew his way around Isle Royale, literally, traveling the shoreline of the entire island when weather permitted. He taught me the waters and hazards in the nearshore areas. He showed me how to read the nautical charts and emphasized the importance of having them with me always. Because fog often covered the lake, he showed me how to navigate by keeping a log when skies were clear so when visibility was poor, I could maintain a compass heading, monitor my speed, and track the time I traveled along that bearing to clear a given reef or island. This was often quite tricky, and an error could cause me to shear a pin on the motor or even crash on a reef. In case of some problem with the motor, I also carried a 10-horsepower outboard as a spare.

I also had to contend with Lake Superior's waves and swells. When the big lake kicked up, which was often, the waves grew to the size of ocean waves. Many were too huge to traverse in my small boat, and I chose not to deal with them. Sometimes the lake was flat calm, and zooming along at top speed was a cinch. Conditions in between were tricky. Were waves too high or not? Sometimes I set out only to think better of my decision and turn back. Other times I simply dealt with the waves the best I could, especially in a following sea. I would ride up the wave to its crest, throttle back, and surf the crest as it broke downward. Then I would throttle up and climb the next wave. A full set of rain gear kept me reasonably dry. But heading into waves of this size was another story, and I never tried to buck them. Rather I would wait for the sea to drop, which sometimes took days. Fortunately, I was quartered on a protected bay, so if the lake was too wild, I could cross the bay, dock the boat, and hike to where I was going.

Bob helped me solve the problem of how to safely stow the boat when I was off exploring the trails for several days. As an experienced Isle Royale navigator, he knew the safe harbors and beaches of the island. He taught me how to moor the boat between two ropes, one tied to a dock of some sort and the other to the shore with the bow pointing out toward the lake. This way there would be enough give in the ropes so the boat could rise and fall with incoming waves. There were only four or five places like this around the island, but there were trails leading away from each of these locations, so I could travel wherever I chose.

With the *Wolf* and my newfound navigational skills, I had the tools I needed to explore the entirety of Isle Royale. I followed the shoreline and navigated the shore to Isle Royale's lighthouses— Rock Harbor near park headquarters, Passage Island on the bleak spit of rock on the island's far northeastern point, and Rock of Ages at the far southwestern tip off Washington Harbor. I found my way to the isolated homes of the old-line commercial fishing families, often to buy fresh lake trout for dinner or to sip a cup of tea at the kitchen table. Because of my lifetime aversion to coffee, it was at first a bit awkward to have to turn down coffee from fisherfolk whose veins were full of it, but to a person, each was always a perfect host. And every week, I would motor across the channel from Mott Island to Rock Harbor for Sunday mass. I was still a believer then and valued attending church in the little makeshift Sunday chapel set up in the rec hall.

As I boated along the southern shore of the island, Lake Superior stretched unbroken to the horizon. But as I turned the corner at Blake Point on Isle Royale's northeastern tip or cleared North Gap at the other end of the island, I saw the gray-green hills of Minnesota's North Shore and the cliffs of the Sleeping Giant off the entrance to what was then called Port Arthur and Fort William, less than 20 miles away.

That 20 miles, more or less, made our project possible. It was close but not too close.

It was close enough that various species, including moose and wolves, could reach the island—but far enough that many species did not. The result was an animal community and ecosystem that resembled that of the nearby mainland but was streamlined in comparison. Fewer players—only one large herbivore (moose) and one large carnivore (wolves)—simplified the complicated dynamics among species, making them easier to study while providing valuable insights into the workings of the more complex mainland systems.

And the distance was great enough to keep all but the most adventurous animals from leaving. So someone like me could be assured that the wolves I saw one day were among the same ones I would see the next, making possible a reasonably accurate census of the large animals on the island. Likewise, if 10 percent of the moose were to disappear from the population, we could surmise that the cause was wolf depredation or other natural mortality and not that large numbers swam off toward Ontario. In that respect, Isle Royale was the perfect island laboratory: large enough to contain a self-sustaining population of moose and wolves that resembled natural populations elsewhere in the north woods; small enough and contained enough to count the number of primary subjects—moose and wolves—with some expectation of accuracy. From the science standpoint, 210-square-mile Isle Royale was the sweet spot between a vast wilderness and a laboratory. We couldn't have designed a better study area.

That Isle Royale did in fact harbor wolves constituted a special advantage: it was one of the few places in the Lower 48 that did. Wolves had once occupied the entire continental United States. But the European settlers began killing wolves virtually from day one, because wolves competed for prey, killing game animals and livestock. Wolves were eradicated by trapping, hunting, poisoning, digging up dens—much of it by government agents. The animals were extirpated even from remote wilderness in the West because of fears they would repopulate cattle and sheep country. So thorough had this persecution been that only a few hundred wolves remained in the Lower 48: a few in northern Wisconsin and the Upper Peninsula of Michigan, but most in northeastern Minnesota. In this way, too, Isle Royale National Park was an anomaly, a place where wolves existed and where questions of mortality wouldn't be complicated by human hunting. With prospects so poor for the wolf over much of its range, it was important that the species be studied while and where it was still possible to do so.

Moose and wolves, by all accounts, were relative newcomers to the island. Moose appeared in the early 1900s as the animals proliferated in Minnesota and Ontario in the wake of logging that created young forests and lots of browse, the tender young twigs moose eat. How did they arrive? They might have crossed the ice on a cold winter, but moose are wary of walking on bare ice, where their footing is poor. They probably swam, as moose are often seen swimming along the mainland and the island. Twenty miles of open water would not be insurmountable. By 1915, there may have been as many as two hundred on the island. By the late 1920s and 1930s, their number had risen to more than one thousand—maybe three thousand. At the time, no one had devised a good way to accurately count them. But moose were plentiful enough that they ate their favorite foods to nubs, and the population collapsed. A vast forest fire swept a quarter of the island in 1936, further depleting moose browse. But with moose numbers already low, the burn and surrounding forest quickly recovered. Moose responded with another population explosion—to about eight hundred, according to estimates at the time. Once again, they ate themselves out of house and home, and their numbers dropped.

Into the midst of this boom and bust appeared the wolf. There had been sporadic reports of wolf tracks in the early 1900s. But clearly by 1949 they had arrived and formed a permanent population. It's possible they had arrived by swimming, just as the moose likely did. But it seemed far more probable they had crossed the ice during a cold winter. Lake Superior freezes over completely on average every twenty years or so, but an ice bridge from the north shore to Isle Royale is a much more common occurrence—in the 1960s, it was the rule. And several times observers on Ontario's Sibley Peninsula had reported seeing tracks, and even the wolves themselves, proceeding from the mainland far out onto the lake.

So for the first time in recent history, moose and wolves lived on Isle Royale together. When I arrived, it had been just ten years or so, and we had no idea how stable this system was going to be. Would the wolves wipe out the moose on the limited confines of the island? Would the moose continue to proliferate and repeat their boom-and-bust cycles? Would the wolves simply one day leave the way they came? Might this drama have played out before—wolves and moose appearing on the island and vanishing, only to reappear centuries later? At the time, we knew the answers to none of these questions.

What did become clear as I talked to old-time residents of the island is that Isle Royale had recently hosted quite a different cast of characters. A century earlier woodland caribou had roamed the island (and were the most common hoofed animal along Lake Superior's northern shore as well). But they seemed to have disappeared from Isle Royale about 1925, said the old commercial fishermen who hunted them, as their range on the mainland was retreating steadily northward. A dozen white-tailed deer, never common in this far-north country, had been introduced in 1906 but died out.

Two fairly large carnivores had also been common once, but they too had disappeared or at the very least had grown mighty scarce. Local fishermen had trapped Canada lynx by the dozens during the very early 1900s, but now lynx were nowhere to be found. Likewise, coyotes (called "brush wolves" in these parts) had somehow found their way to the island by the early 1900s but, like the lynx, had virtually vanished. Perhaps that was the work of the wolves. Large canids are intense rivals for food and space, and on the limited territory of an island, even an island as big as Isle Royale, it was conceivable the wolves had killed every last coyote.

Just as a select few large herbivores and carnivores found their way to Isle Royale, so did a number of smaller mammals common to the nearby mainland. Beavers, red foxes, snowshoe hares, red squirrels, mink, weasels, muskrats, deer mice, and various species of bats had all made the trip at one time or another. Some were now quite common. How did they get there? No one knew. Clearly the semiaquatic beavers and mink might have swum. Mice may have stowed away aboard boats. But snowshoe hares and red squirrels? It was hard to imagine they had either swum or crossed 15 miles of lake ice without being picked off by a fox or bird of prey.

Not surprisingly, birds common on the nearby mainland were well represented on Isle Royale. We saw bald eagles, loons, and ravens. Various species of ducks dabbled and dove along the shoreline and interior lakes. Hawks and owls patrolled the forest. Warblers and sparrows flitted through the underbrush. All the woodpeckers common to the mainland hammered away at tree trunks in search of insects.

Yet, two iconic birds of the north woods, the ruffed grouse and spruce grouse, had never been reported on the island. These grouse, notably short-range fliers, had apparently never flown to the island. Yet their cousin, the sharp-tailed grouse, had made the flight. A bird of brushlands, the sharptail is scarce in northern

Minnesota and Ontario, but as I walked the trails of the island, I began spotting them, especially in the aspen scrub growing in the wake of the 1936 forest fire.

Likewise, many of the other critters you would expect to see in nearby Minnesota or Ontario were not to be found. There were no bears, skunks, bobcats, badgers, or porcupines. The happenstance of geography—some 20 miles of open water that froze over only occasionally—had set up an interesting natural experiment, allowing a few animals to cross but keeping many others at bay. It was just the kind of laboratory that Dr. Allen had been yearning for.

In retrospect, I can appreciate how little we knew of wolves and how few "wolf experts" there were.

We really didn't know what constituted a "wolf pack" or how packs formed. Were they families? Did unrelated wolves join a pack and engage in a struggle for dominance? What animals did they hunt? Clearly they killed big animals, such as moose, but to what extent did they rely on smaller critters such as hares? And what effect did wolves have on the animals they hunted? How often did they cull the sick, old, and injured? Or was any animal fair game? At the time, the wolf was still viewed as evil incarnate, a marauding machine that could kill at will, taking whatever animal it wanted and invariably decimating herds of deer, elk, and moose (an attitude that would persist among ungulate biologists for years).

The ad hoc advisory committee Dr. Allen had assembled to meet with me when I arrived were great guys who enjoyed sharing outdoor stories over cocktails, but there wasn't much wolf expertise among them. Dr. Allen himself had never studied wolves. Larry Krefting, a researcher with the U.S. Fish and Wildlife Service, had visited Isle Royale over several years to study the effects of moose browsing on the forest plants but had never studied the relationship between moose and their only predator on the island.

Of the committee members, only Milt Stenlund, a wildlife biologist with the Minnesota Department of Natural Resources, and Douglas Pimlott, a Canadian wildlife biologist, had really studied wolves in a scientific way. Stenlund had tracked wolves and spotted kills in northern Minnesota by airplane in winter to gauge the dynamic between wolves and white-tailed deer. That experience was of considerable relevance to my upcoming fieldwork, and I was glad to have Stenlund in my corner. Pimlott was just getting started with his wolf research and would go on to lead several Canadian environmental organizations, write *The Ecology of the Timber Wolf in Algonquin Provincial Park*, and start the International Union for Conservation of Nature's Wolf Specialist Group. I would continue to correspond with him in the years ahead and visited him once in Algonquin Park. But at the time I first met him, his only experience with wolves was keeping a captive colony of them in Ontario.

Perhaps the person whose work was most relevant to my project was James Cole, a National Park Service biologist who began observing and counting wolves on the island beginning in the early 1950s. Cole had collected and analyzed wolf scats and had even flown over the island in winter to conduct a census of the moose and wolves in the park. He had really pioneered some of the techniques I would be attempting in my three years here. I had access to his typewritten reports, but unfortunately I never had the chance to meet the man.

The published literature was similarly sparse. Sigurd Olson, who would go on to become a popular outdoor essayist and conservationist, had studied wolves in northern Minnesota in the late 1930s. Ian McTaggart-Cowan, head of the zoology department at the University of British Columbia, had studied wolves in national parks in the Canadian Rockies and their reliance on elk, mule deer, and to a lesser extent snowshoe hares and beavers. Cowan's study, published in 1947, was striking for its conclusion that "wolves are not detrimental to the park game herds, that their influence is definitely secondary, in the survival of game, to the welfare factors, of which the absence of sufficient suitable winter forage is the most important."

Alaska biologist Bob Burkholder was tracking wolves across

vast reaches by plane about the same time I was starting my work on Isle Royale. Lois Crisler, though not a trained wildlife biologist, was writing perceptive observations of wild and captive wolves at about the same time.

If one work stood out as meaningful—even inspirational—it was Adolph Murie's *Wolves of Mount McKinley*. Murie, a wildlife biologist for the National Park Service, was the first scientist who really met wolves where they lived, spending weeks and even months in the field and logging hundreds of miles on foot or by dogsled. Murie, who grew up in Moorhead, Minnesota, and graduated from Concordia College and the University of Michigan, had actually studied moose on Isle Royale in the 1930s before leaving for Mount McKinley (now Denali) National Park in Alaska between 1939 and 1941, where he studied the relationship between wolves, Dall sheep, caribou, and moose. He sat within view of wolf dens—one time for thirty-three hours continuously—to study wolf behavior. He once even climbed inside a wolf den. In the wide-open landscape he watched wolves stalk and chase down prey, testing the herd for weak animals. He examined wolf kills and concluded "that wolves prey mainly on the weak classes of sheep, that is, the old, the diseased, and the young in their first year."

Reading *Wolves of Mount McKinley* only stoked my boyhood fantasies. Adolph Murie had actually done what I, in my wildest dreams, aspired to do.

Unfortunately, my work this summer was much less exciting. Resigned to not seeing wolves with the forest in full foliage, I spent my days walking trails from one end of the island to the other, often sleeping out in the various campers' lean-tos scattered about the island. As I walked, I noted locations of wolf tracks and then cleared the trail by erasing the tracks with my foot so that I could distinguish new tracks the next time I passed by. However, there was no reason to think that tracks would tell me anything other than that wolves were there and traveling the same trails with some regularity.

I collected wolf scats, saving them in paper bags, labeling them with date and location—like a dog owner cleaning up after his pet but with greater fastidiousness. I slipped the bags into a side pack and took them to an outbuilding, where I opened the bags and dried them out. My place was littered with wolf shit.

But scats were valuable. They revealed what percent beaver and what percent moose made up the wolves' diet. They also told me what percentage of the wolves' diet was calves and what percentage adults.

People working on the island had learned of my project and often relayed reports of wolves. Soon after I arrived, a Wisconsin tourist told rangers he had spotted six grown wolves on Huginnin Cove Trail at the other end of the island. I quickly received word and laid plans to travel out there. But, really, what good would it do? Sightings were even more ephemeral than tracks.

So eager was I to actually see the animal I was supposed to be studying that I occasionally buried beef bones and meat scraps from the kitchen at various sites around the island. One day I was checking on just such a cache on Siskiwit Beach near the mouth of the Siskiwit River. Tracks revealed a wolf had emerged from the woods and walked about 150 yards along the beach directly to the bait. It had dug up most of the scraps and pawed beneath a big femur but had left it in the sand. The animal had left tracks of such freshness and clarity in the wet sand that I was motivated to rebait the hole and build a blind nearby to spend the night in hopes it might come back. A thunderstorm postponed my plans for a day, but the following night I sat, watching from my blind. I waited from seven until midnight, but no wolf appeared.

The closest I came to seeing a wolf that first summer occurred one evening in mid-July. I had spent the day collecting scats and clearing trails and hiked in to the Hatchet Lake patrol cabin, where I saw a cow and calf in the water at the east end of the lake. Later, as shadows lengthened, I heard a long, mournful sound. Yet I was still so green I couldn't say for sure if it was a moose or a wolf. Only later, when I told Bob Linn about the experience, did he say it was most likely a wolf.

As I walked, I often saw moose. Early on, I approached one bull with my camera and got to within about 25 feet before he spooked. Seeing moose was exciting, especially at first, but really of little consequence. We knew moose roamed all over the island. True, we could gauge the bull-to-cow ratio and the cow-to-calf ratio, and get some idea of the habitat and areas of the island that moose preferred. But the sample was small, and it would not be until I could search the island by plane that I would see enough moose to collect really useful information.

But moose mandibles were another thing, and I collected them as I found them. The heavy durable jawbone could tell some interesting stories. The wear on the teeth gave me a good idea of the age of the moose when it died. Also, some moose would get "lumpy jaw," a debilitating necrosis more technically known as actinomycosis that typically began at the first molar and ate deep into the gum and bone of the lower jaw. The effects of this bacterial infection were clearly visible in the scavenged mandible and suggested the moose was probably malnourished and potentially vulnerable to either starvation or a wolf attack. And marks on the bone could suggest whether wolves had killed the moose or scavenged the carcass. I sent out word to other people on the island to bring me moose mandibles or whatever other remains they found. One of my first discoveries came when I hitched a ride aboard the tugboat *Tonawanda* to Todd Harbor, where the construction foreman showed me the remains of a moose calf—a couple of mandibles, a shoulder blade, rib, other bone shards, and scattered hair. As it turned out, hanging out with the construction crew was a handy way to get around the island. Often I would hike across the island and then catch a ride back to headquarters with the guys.

Such was my first summer. By mid-August, I had still not seen a wolf. But I had learned the geography of the island and met many of the people I would be working with for the next three years. I planned for winter use of the lean-tos and cabins in the backcountry and discussed with Bob Linn my plans for the winter. I made an inventory of supplies and began gathering what I would need. Dr. Allen came for a visit, and I took a boat with him, his wife, and his daughter one last time to Daisy Farm Campground, not far from headquarters. Then I packed up my personal belongings and boarded the *Ranger II* at the dock at Rock Harbor and set out across the wide expanse of Lake Superior back to the mainland, civilization, an impending marriage, and a full load of fall graduate courses at Purdue. My thoughts, though, were of my return in February, when I would begin counting the wolves and moose.

EYE IN THE SKY

WINTER 1959

❧

We spent the morning at the airport in Eveleth, on Minnesota's Iron Range, hauling gear to the two light planes parked near the runway: a Cessna 180, a high-wing single-engine all-purpose cargo hauler popular as a bush plane, and a much smaller Aeronca Champion, the plane we would fly each day during our Isle Royale aerial surveys.

It was February 4, 1959, typical winter in northern Minnesota—cold, snow plowed up along the perimeter of the airport and lying deep in the woods beyond. We lugged duffels, packs, boxes of food and equipment, and miscellaneous tools and gear until both planes were stuffed to the ceiling. Art Tomes, the president of Northeast Airways, based in Eveleth, had taken a personal interest in our wolf project and would fly the Cessna. Jack Burgess, who would stay on the island for a week and fly the first few surveys, piloted the Champ. At 10:50, after a last-minute check to make sure we had clear weather for our flight to the island, we jammed ourselves into the planes. I flew with Frank Taddeucci, a Park Service mechanic, in the Cessna. Bob Linn, the park naturalist, flew with Jack in the Champ. Our planes rumbled down the runway, picked up speed, and labored into the sky.

We gained altitude as we flew across northeastern Minnesota and followed the gnarled finger of Pigeon Point into Lake Superior. I was told we wanted to reach at least 8,000 feet so that if an engine conked out over open water, we would have a chance to glide either to the island or to turn and sail back to the mainland. Not exactly a confidence builder. From our height I could see that except for in the protected harbor of Port Arthur and Fort William, there was very little ice on the open waters of the big lake. To put down in Lake Superior, even successfully, would be a death sentence.

As we crossed from the mainland over the lake, we could see the entirety of Isle Royale. It was a shock and exciting besides. Unlike approaching by boat, when the island slowly emerged from the horizon until it appeared a solid wall of rocks and trees, I could see the whole thing—Windigo and Washington Harbor in the foreground, the rest of the island stretching out to the northeast, the rocky ridges and harbors reaching into the distant lake. I had memorized that landscape from my summer hiking, and here it was, laid out like a map, surrounded by water. It truly was a world of its own.

In just a matter of minutes we nosed down over Washington Harbor. I spotted three moose along the shore and snowshoe hare runways in a spruce swamp. The ice in the protected bay was covered with wind-driven snowdrifts, and as we set down, the skis thudded and the plane shook. We taxied to the ranger station, ranger Dave Stimson's summer home, which had been shut down since early autumn. The park staff and even the island's resident fishing families had decamped for the mainland by early November, before the possibility of a freeze-up that would make boat travel impossible.

We clambered out and immediately began unpacking the gear we had stowed only two hours earlier. We lugged some through the snow to the ranger's cabin by hand. Some we loaded aboard the Studebaker Weasel, a military-surplus tracked vehicle. We packed everything into the cabin where we would spend the next six weeks. Frank Taddeucci fired up the generator so we would have electricity. We spent the next few hours checking radio lines, heating the transmitter shack, and freezing anchors into the ice to tie down the Champ so it wouldn't blow away. In midafternoon, Frank and Art Tomes climbed aboard the Cessna, roared across the ice, and lifted off back to Eveleth.

The oil heater slowly warmed the frigid cabin. We settled into our new surroundings and unpacked and arranged the equipment we would need for our surveys. Because of the cold, the plumbing was shut down and drained. But we had an outhouse in the woods behind the cabin, and we chopped a hole in the Lake Superior ice by the dock and carried water back to the house in a steel milk can. The temperature dropped to minus 5 degrees that night, and we would have to chop fresh ice from the hole each time we drew more water.

I was leaving behind not only my classmates and colleagues at Purdue but also my new wife. Betty Ann and I had met and got engaged while we both attended Cornell, where she was studying entomology. We had married late the previous summer on Long Island, where her folks lived. It was a low-key wedding, with the reception in her parents' garden and a barn her dad had remodeled. We both headed for Purdue for fall semester. Betty Ann was pregnant now and staying with her folks on Long Island while I was on Isle Royale for the six weeks that would either make or break my plans to become a wildlife biologist.

The next day we began hauling canned goods from our root cellar into the now-warm house (though it would be a few days more before the houseflies from last summer would find new life and begin to buzz around their newly habitable surroundings). Bob Linn checked out the equipment in the station's radio room and climbed up to the radio tower so we would have communications with the mainland. I helped Jack with the plane. We assembled an emergency kit of sleeping bags we packed in the cockpit, and snowshoes we lashed to the wing struts. Bob, Jack, and I were the only people on the island, far from other humans. If we had an emergency, we were on our own.

The following morning was cold: minus 13 degrees, with flurries dry and light, like icy dust. We still weren't ready to fly. So after breakfast, I pulled on wool pants and a coat and set out on snowshoes along the Huginnin Cove Trail, which led from the ranger station up to the north shore of the island and back in a 9-mile loop. Near the footbridge over Washington Creek I found an

41

unusual animal track in the snow and followed it into the spruce swamp south of the stream. The tracks were old and had filled a bit with blown snow, but I could still measure them: 3½ inches across and 4 inches long. They were the size of wolf prints, but as I followed them for about a mile, I realized the animal exhibited very unwolflike behavior.

The trail zigzagged through the conifers. The animal, whatever it was, had jumped up on logs and blown-down trees, and even occasionally onto branches. For more than 100 yards it had walked almost entirely on blowdown, only occasionally touching the ground. I already knew enough about canids to realize that a wolf wouldn't engage in or even be capable of such acrobatics. Considering the animal's agility, I figured the tracks must belong to a cat. And the only cat with paws that size that had ever been reported on Isle Royale was a Canada lynx. I was excited. Not only was the lynx a large and charismatic icon of the north woods, but it also had gone unreported since many were trapped in the early 1930s. As I followed the trail, I also spotted tracks of two weasels, several mice, red squirrels, and—of greatest interest to any lynx that might be on the island—snowshoe hares, the lynx's almost-exclusive prey.

We slept in the next morning until eight forty-five, which is surprising in retrospect, since, weather permitting, this was the day it would all begin. We planned to fly our first reconnaissance of the island to look for moose and, especially, wolves.

The day was overcast but the clouds were high—good enough to fly, Jack Burgess decided. So he and I prepped the Aeronca Champion. I had flown in a light plane only a few times—three days ago on our way to the island, and a couple of times as a wildlife student at Cornell when a student pilot took several of us up one at a time to observe and count a deer herd that yarded up during the winter on the south-facing slope of Turkey Hill, near campus.

The Champ weighed only a bit more than 700 pounds, less than half the weight of a Volkswagen Beetle. It seemed flimsy, too insubstantial to carry two of us aloft. A high wing and capacious cockpit (for such a small plane) with a lot of glass and a prominent windshield gave it the appearance of a cartoonish mechanical drag-onfly. It cruised at 80 miles an hour and would slow down to less than 40 before stalling. It carried enough gas for two and a half hours aloft. To start it, Jack or I would grab the propeller, swing a leg back, and give the prop a good wrench downward. Eventually, the engine would fire, the prop would spin, and we were ready to fly. I tucked into the back seat, behind Jack, and we sped down the bumpy snow-covered ice, the engine pounding, until suddenly we jumped into the air and climbed above the island.

It was barely above zero outside and not much warmer in the plane. The air was "bumpy," and the light plane lurched in the wind. As we hurtled through the sky only a few hundred feet above the bare trees and snow-covered ridges, I looked at the map of the island in my lap and jotted notes on my clipboard. We flew northeast, along the length of the island, sticking closer to the south shore. We quickly began seeing moose in the snow below. Many were bedded down but rose to their feet as the plane passed overhead. Along the south shore of Siskiwit Lake, we spotted a red fox. We swooped down to within 100 feet, but it remained curled up and simply watched the plane. I spotted a sharp-tailed grouse, and Jack saw three more.

Suddenly, as we spiraled above the trees with my face either buried in my notes or pressed against the side window, I felt a thick sour taste rising in my throat. Jack recognized immediately what was going on and directed me to the empty coffee can he kept for his passengers, and soon I was filling the can with my breakfast. If only the nausea had gone away after vomiting. But it didn't, even with nothing left in my stomach. We headed back toward the ranger station. I recorded the locations of nine moose in all but didn't feel any the better for it until, barely more than an hour after we took off, the skis hit the snow, we slid to a stop, and I tumbled out of the plane.

That afternoon, I ate a big lunch, as if to replenish everything I had lost, and took a Dramamine.

By 3:20, we were back in the air. Through the afternoon, I felt sick but managed to keep my lunch down. The sky had cleared, and I spotted several moose where the sun brightened the snow—a cow and calf south of Lake Harvey, six moose together on a point at the east end of Todd Cove, a lone moose on a steep ridge along the north shore of the island, and another in the shadows along the shoreline. Tracks were abundant off the north shore of Lake Halloran near the island's south shore.

In all, we spotted seventeen moose, which I suppose confirmed the validity of our methods for at least half the project. But despite having flown much of the island, we had not seen a wolf. And despite holding down my lunch through the afternoon, I wondered how I would manage to finish this project if I turned green every time I flew.

The clouds hung low the next morning, and we put off our flight until shortly after noon.

As I flew behind Jack Burgess, I quickly realized what a joy it was to watch him handle the airplane. He would power up over a ridge and glide over the top. Everything was smooth. There was no wasted motion, no wasted power. He was a small, slight man, and as I watched him, I thought of a jockey on a prize-winning thoroughbred.

Jack was from Tower, in northern Minnesota. Back then it was still legal to hunt wolves for bounties in Minnesota, and Jack was an aerial wolf hunter, piloting the plane as a hunter rode shotgun, literally. As Jack spotted a wolf in the open, he would swoop low over the running animal while the hunter shot it with a rifle or shotgun loaded with buckshot. He told me that he had once spotted a pack on frozen Lake Vermilion and swooped low to harass them—so low he nearly skimmed the snow. One of the wolves, Jack swore, leaped up, bit the wing strut, and pulled the plane into the snow. Jack survived the crash but had to walk back to a road to get help. The story impressed me immediately, and with time I began to appreciate the irony that my field study on the wolves of Isle Royale was being enabled by a wolf hunter.

Despite Jack's experience in spotting wolves from a plane, we both struggled to identify their tracks—or what might be their tracks—in the snow several hundred feet below. Often we would spot a line of tracks winding among the birch, spruce, and fir, but without a reference for scale, we had no idea what they were. At one point, as ridiculous as it is to admit now, I thought the long strings of tracks along the rocky shoreline of Isle Royale must be mink tracks. I had trapped mink and was familiar with their tracks and how they tended to wend along the shoreline for great distances. But I had no idea what their tracks would look like from an airplane. Finally, as we spotted more moose and the tracks they left behind, we gained a better sense of scale, and we began to realize we were looking at wolf tracks. And there were a lot of them.

Passing over Tobin Harbor, a long, narrow bay on the island's northeastern end, we spotted what we now felt sure were wolf tracks along the north shore of the harbor. They seemed to be concentrated along a small point nearly directly opposite Rock Harbor Lodge. As we circled above, we eventually saw what appeared to be a wolf-killed moose on the snow. We nosed down, and Jack landed the plane on the ice near the carcass. I knew that finding these kills would be an important part of my work. Adolph Murie had discovered in Alaska that wolves tended to kill the young, the old, and the debilitated. So it was important for me to determine the age and condition of any dead moose we saw. I climbed out of the plane and hiked to the carcass.

This was quite a thrill for me: my first wolf-killed moose! I had seen many coyote-killed deer during my Adirondack Mountain deeryard checks. But this was much different. The remains of the dark moose lay scattered on a track-packed, blood- and gut-stained patch of snow half the size of a basketball court. The freeze-dried hide was wide-open, the long backbone with head attached lying alongside, ribs well chewed to the nubs, and a big chunk of leftover stomach contents frozen to the ground. Well-chewed leg bones barren down to the knees showed how thoroughly the wolves had eaten this unlucky beast. And widespread wolf scats, white raven shit, and more dirty blood and frozen body juices colored the scene of the wolves' feast. I was impressed.

I quickly realized that it would be easy to distinguish a moose the wolves had killed from a moose that had died of natural causes that they had merely scavenged. The multitude of tracks and volume of fresh blood told the tale. I chopped off the skull, including the mandible. The teeth would give me a good idea of the moose's age, though I could easily see that this was a calf. The condition of the mandible might help me gauge the health of the animal. I pulled out a hacksaw and cut through the animal's femur to examine the bone marrow. Marrow depleted of fat was a sure sign that the animal was malnourished, either because of lack of food or parasites or other illness. My experience as a Cornell student examining bone marrow of starving deer on Turkey Hill and in Adirondack Mountain deeryards would pay off here.

We took to the air again, but our day was cut short as low clouds moved in. We had flown less than three hours, but we had spotted twenty-two moose, singly and in small groups, including a cow with two calves. I had examined my first wolf kill. And once we realized what we were looking at, we had begun to see wolf tracks everywhere. Perhaps best of all: I didn't get sick.

Snow fell the next day during breakfast. But it didn't last long, and the clouds began to break, so Jack and I took off in the Champ and flew along the north shore of the island toward the dead moose on Tobin Harbor. We spotted a live moose along the shoreline. And then Jack called my attention to an animal on the ice near the east end of Todd Harbor.

It was a wolf—finally! It lay on a large chunk of ice about 100 feet from a point along the shore. We circled and came in low to take photos. That was enough for the wolf, which sprang to its feet and ran back up its old trail and hid beneath overhanging ice along the shoreline. But our wolf count had started.

Still, it was a full pack that I longed to see. We continued east, where we spotted wolf tracks weaving through the snow at the mouth of McCargoe Cove, a deep notch cut into the island's north shore. Jack banked and turned down the cove, where we spotted four wolves on the ice—and two more lying on the bank! The wolves stood and ran around on the snow-covered ice, slipping as they—inexplicably—tried to run *after* the plane.

From the air, we began to follow their tracks back from where the wolves had come. We began seeing more tracks on Amygdaloid Island and soon spotted a pack of nine wolves on a ridge along the south shore of the island. As we circled, some of the wolves remained curled up on the snow, while others ran back and forth, as though they were excited but didn't know how to respond to an airplane. We came in low over a wolf, and, again, it chased the plane. As we climbed away, the wolves rolled in the snow, stood, and shook off the loose snow.

Fifteen wolves, more or less in one place! That was a big pack and certainly more than I ever expected to see at one time. In a way, seeing those wolves validated the entire project. I never really doubted that we would be able to find wolves and follow them in the plane, but until it actually happened, it remained an open question. Now that question was answered. I had seen my first wolf—and then some. And it was clear that we would be able to find, watch, and follow them. That day, February 9, in many ways was the day my project began.

The following day we were grounded by snow. Just when we were getting started! I sat in the cabin, writing letters to go out in the next day's mail aboard the Cessna coming from the mainland. Toward the end of the day, I hiked out to the spruce swamp along Washington Creek, hoping to see more unusual animal trails, but high winds and fine snow had filled the tracks, making them unreadable.

The next day, too, was bad, and we decided to stay in. However, the Cessna arrived from Eveleth, bringing with it a new pilot. Jack Burgess would rotate out, while Don Murray, a big, gruff, bearded fellow, would fly our surveys for the next three weeks. The plane also brought a reporter from the *St. Paul Pioneer Press*, who interviewed me about our wolf project. I was glad I no longer had to admit I had never seen a wild wolf.

Don was raring to go, and I liked that. I had no idea how he would work as my new aerial chauffeur or whether he would end up flying me during my next two winters. But Don seemed as interested as I was in finding the wolves. After checking the Champ, we lifted off and headed over the north shore of the island. At Todd Harbor, again, we spotted fifteen wolves in a group. We backtracked their trail in search of a kill and then returned in search of the wolves again. We found them at Wilson Point, at the west end of the harbor—a group of eleven walking the southern shoreline of the point in single file, and the remaining four on the north side, lying on the lake ice.

From that day on, weather permitting, we were able to find, track, and follow the pack of what now appeared to be fifteen wolves almost at will. Even after a day of bad weather, we knew where we had left them and could pick up their trail and usually find them. Over time, we began to understand where and by what routes the wolves usually traveled, which made finding them again much easier.

One day in mid-February Don and I lifted off on a calm, sunny morning and soon found tracks near the mouth of Washington Harbor, only about 3 miles west of our camp. We followed the trail backward to Huginnin Cove. Circling back toward Washington Harbor, we spotted the large pack north of Thompson Island, just off the southwest shore of Isle Royale. Eleven were walking on the ice, and another stood on a ridge nearby. Circling the pack several times, we spotted a dead moose on the island. Don and I decided to set down on Washington Harbor so I could hike in to examine the kill.

Bird tracks mingled with wolf tracks showed that ravens, as well as wolves, had been feeding on this carcass. I sawed through the femur to examine the marrow and grabbed the mandible. I could see all fifteen wolves sprawled on the ice southwest of Thompson Island, only about 300 yards away. I could tell they were watching me, but they seemed unperturbed. Three ravens walked among them. I hiked back to the plane. We lifted off and flew over the wolves at an altitude of only about 150 feet. Still the wolves seemed unconcerned. One even rolled in the snow.

Flying along the south shore of Isle Royale, we found fresh tracks of what appeared to be a single wolf near an old kill in Tobin Harbor, at the far northeastern end of the island. If indeed the track was fresh, it suggested that at least one other wolf lived far from the pack of fifteen we had just watched 40 miles away at Washington Harbor.

I still admired Jack Burgess's elegance at the controls of the Champ. Don was a bigger fellow. In contrast to Jack with his nimble physique, Don seemed as if he had been stuffed into the cockpit. But I quickly came to really like him, as a pilot and as a fellow researcher.

Despite his mountain-man appearance, he was easy to get along with in camp. He was fond of saying, "I've got to get my beauty rest," which was, to anyone who knew him, the height of irony. He was the mechanical wizard of our crew. Not only did he keep the plane flying, but he also could repair small engines and get them to start in the subfreezing temperatures. He had run a motor pool during the Korean War, and his combination of organization and fix-it skills showed in his work. He loved the outdoors. He hunted and fished. In fact, when I hiked off to inspect moose carcasses, I would occasionally return to find him fishing through a hole he had drilled in the ice.

Like Jack, Don was an aerial wolf hunter in northern Minnesota. When he started on our project, he really didn't like wolves, especially later when he finally saw them chasing and killing moose. He compared the wolves hanging on a moose to the lampreys that were clinging to Lake Superior's lake trout and decimating them. But eventually, as he saw the great distances wolves had to travel and how difficult it was for them to actually catch and kill moose, he seemed to find some sympathy for them and even began to root for them. (In fact, as time went on, Don visited local schools with color slides of the project and argued for more protection for wolves in northeastern Minnesota.) I felt comfortable in the plane with Don at the controls. He was a good pilot. He seemed to have a knack for circling a target, whether it was a moose or wolf, and keeping it in view, despite the trees, the rough terrain, and the frequent buffeting of the wind.

Which is not to say everything went smoothly. Once, as Don was flying to refuel at a gas cache we kept at park headquarters at Rock Harbor, he seemed to get lost in a daydream and flew right by it, overshooting it by several miles. Another time, as we circled the island in search of wolves, the Champ's cabin suddenly began to fill with smoke. Don immediately dropped the plane down at Robinson Bay. He quickly discovered the problem. The oil-filler cap was loose, and oil had spilled all over the hot engine. He tightened the cap, and we took to the air again.

Despite a few moments of anxiety, those were small things. We never talked about the possibility of crashing or making a forced landing and being unable to take off again, though we did carry the sleeping bags for just such a contingency. Getting stuck out on the island could be serious business. While Bob Linn and any other Park Service employee who was staying with us at the ranger cabin had radio contact with the mainland, we in the plane had no radio communication at all—either with the ranger station or the mainland. Help would be a long time in coming.

Don and I had been flying over several parts of Isle Royale in late February when we spotted the large pack walking on the ice along the south shore of the island. We checked on them through the morning as they moved northeastward along the shore. After about three hours, they had traveled about 10 miles and had reached McCormick Rocks, a reef about a half mile off the south shore of the big island. They were headed northeast toward Fisherman's Home, where commercial fisherman Sam Rude and his family lived on a small cove during the open-water season. By now, we knew how the wolves traveled this area, cutting inland some 100 yards from the shore to the head of the cove, then northeast along the cove right by Sam's fish house. That gave me an idea.

"Don," I shouted over the engine's drone. "Let's ambush these wolves. We can land on the cove, duck into the fish house, and watch the wolves go by, maybe get some pictures."

Don was all for it. He deftly landed on the ice near Fisherman's Home, still at least 2 miles ahead of the wolves. He left me and flew off again to pick up Bob Linn to photograph the wolves from the air. I figured that if the wolves followed their usual routine, I might have a chance to photograph them as they walked by. At the least I would be able to get a good look at them.

Sam Rude's place sat on the small point that sheltered Fisherman's Home Cove from the big lake. Houghton Ridge, a rocky 300-foot-high hill, rose behind the homestead. It was a picturesque spot, the house and various other buildings oriented to the cove. At the end of a short dock sat a tar-paper shack. This was Rude's fish house, where fish were cleaned and packed in ice cut from the lake in early spring and stored until a boat could pick them up and take them to market. This time of year, the fish house was empty, or nearly so, and the door was unlocked. So I crept inside, left the door ajar, and kept an eye on the shoreline.

After dropping Bob off at the ranger station, Don returned to Fisherman's Home, and I heard the plane settle onto the ice around the backside of the point where the wolves wouldn't see it. Don hurried over to join me in the fish house.

"The wolves are just up the shore," Don whispered. "I wasn't sure I'd make it back here in time. I hope they come through."

Only a few minutes later, about 4:30, as the sun was swinging low, the wolves appeared. First one, then another, then a few more filed out from the brush and onto the ice. Single file was their usual pattern of travel. It no doubt helped them trench through the deep snow in the woods, but they traveled that way even along the ice-covered, rocky shore.

Out came the rest of the pack. They seemed almost relieved when they hit the smooth, even surface of the cove, hesitating, resting, and interacting a bit with each other as they more slowly ambled ahead. I often noticed this kind of playfulness when the wolves reached an open stretch of snowy ice. Don and I were staring down the bay through the partly open door, eagerly watching this canine performance and waiting for our wild charges to come closer. Observing the wolves from the air was one thing, and we loved every minute of it. But seeing them from the ground, this close, was exhilarating. Maybe I could get a few good pictures.

As the pack straggled by, some 50 feet away, some must have sensed us or our tracks, or perhaps the open door. They had streamed by here many times and no doubt knew the area well. Buildings dotted the island in only a few places, and our shack stood just feet from their usual winter route.

The wolves were curious, and one walked tentatively to within about 15 feet of the open door and peered straight at us. I got a good shot of him (or her), and it then stood broadside for another. A second wolf crossed our tracks on the ice and followed them toward the house. Several turned and came back to the wolves that had come to the building. By my best count, there were at least a dozen wolves, maybe a couple more. But they didn't linger. They continued along the shore and passed to where we could no longer see them. To get a better look, I stepped through the door and stood out on the dock. One wolf that had apparently lagged behind the pack ran back along its trail when it saw me. Then it doubled back to try to catch up to the pack, passing only about 75 feet in front of us. As I jumped off the dock onto the ice and snow, the wolf turned and ran into the woods. The rest of the pack, more distant, showed signs of apprehension and slowly tracked away from us. They didn't run fast and seemed more curious than afraid.

As the wolves passed out of sight, Don and I hiked up the shore to the plane and took off to follow them. We caught up to them near Houghton Point as they were crossing the ice on Siskiwit Bay toward Crow Point.

As the wolves had strolled in front of us at Fisherman's Home and now again as we watched from the air, I noticed that a large wolf and a smaller one kept close to one another. The larger wolf walked either right next to the smaller one or trailed slightly at her hip, as though walking at heel. The larger wolf tried several times to mount the small one, but she snapped at him and rebuffed him each time.

I did see one mating, but from what I could tell, it happened between two other wolves. When one wolf mounted the other, the whole pack, strung out more than 100 yards ahead of the pair, raced back to the copulating wolves for several seconds of energetic milling about. The mating pair were "hung up." (As with dogs, wolves experience a *copulatory tie*, a combination of swelling of the penis and squeezing by the female's vagina that may keep the pair locked together for several minutes to more than a half hour.) The pair was locked and lying down, back to back. The other wolves seemed to lose interest and went about their travels. When we flew over, the coupled wolves stood and snapped at each other. After about fifteen minutes, they were able to separate and raced to catch up to the pack.

As we followed the pack, there were several other attempts at copulation, but each effort was thwarted as the smaller female sat with her tail between her legs. With each attempt, the nearby wolves rushed to the pair. There appeared to be three female wolves in estrus, ready to breed.

If I seemed preoccupied with wolves' mating habits, it's because nothing much had been reported and written about their mating behavior in the wild. From the standpoint of gathering data, this was good stuff, more valuable natural history information from our novel approach of following wolves from the air.

A couple days after watching the pack up close at Fisherman's Home, I noticed for the first time a lone wolf—not a wolf that was simply *alone* but a wolf that gave all appearances of wanting to hang with the pack but was shunned.

The phrase "lone wolf" has a history going back centuries, suggesting an animal (or a person) who prefers to go it alone. But wolves (like people) are social animals, and going it alone probably is never a preferred option, at least among wolves. Something has probably caused an animal to be expelled from a pack or blocked in its efforts to join a pack. Even at the start of my research, we suspected being a lone wolf is not something the animal chooses for itself.

53

But as with most everything else, we had few if any scientific observations and certainly no insight into what might cause a wolf to live a solitary life. Adolph Murie in his study of the wolves in Denali National Park had not said much about them. Stenlund had spotted many lone wolves each year in his Superior National Forest study but had written only a few sentences about them. Thus one afternoon in late February, when we had our first inkling that we had found such an animal on Isle Royale, I was intrigued.

We had spotted three wolves between Five Finger Bay and Duncan Bay at the far northeastern end of the island. Two were large, one was small, and though there were no obvious attempts at copulation, there seemed to be some sexual tending activity between the smallest wolf (probably a female) and at least one of the larger wolves.

We flew on and soon spotted the large pack more than 2 miles away. The distance and our quick flight precluded any chance that we had double-counted any wolves. That meant that at the very least there were at least nineteen wolves on the island.

The big pack was also one member larger than we had seen before, and we soon noticed that one of the wolves followed about 100 yards behind the main pack and appeared afraid of the other wolves. It slinked along with its head low, ears back, and tail between its legs—signs of fear or apprehension among wolves and dogs alike. We didn't know quite what to make of this wolf, but I made notes of its behavior.

Meanwhile, the remainder of the group filed through the snow. We followed them around the west end of Mott Island, across Rock Harbor about 300 to 400 yards to the remains of a dead moose. One wolf was clearly in the lead, ranging about 150 yards ahead of the others. The status of this "leader" wasn't clear to me. Did it rank high in the pack's social hierarchy? Or was it simply leading the hunt that day? Or did it just find itself eager to go and at the head of the line? At any rate, the other wolves would rest when the leader rested and walk when it walked, except during mating activity, when the lead wolf patiently waited for the others. It seemed somewhat incongruous that a high-ranking wolf—if indeed that's what this leader was—would remain completely aloof from mating activity in the pack.

The next day, we found the wolves again—the pack of fifteen with the lone wolf slinking behind—at the west end of Rock Harbor, a large sheltered bay called Moskey Basin. The pack headed up an old wolf trail to tiny Wallace Lake and then southeast along the ridges, wallowing through drifted snow. Five of the wolves lay down on one of the ridges, while the rest, including the straggler wolf, continued to Lake Richie. They were filing across the lake ice when, suddenly, they broke out of their line and pointed upwind for several seconds. Wagging tails, they regrouped and headed into the wind, single file, directly toward two adult moose feeding a quarter mile away.

When the pack was within 200 yards, the moose took off. One headed toward Lake Richie, the other in the opposite direction. The wolves chased the second animal through deep snow. Soon all but the first wolf gave up. This wolf closed to within 50 feet of the moose and then stopped. The remaining wolves rested 100 yards behind. Meanwhile, the other moose stood 150 yards away, nose upwind, between the wolves and the lake. The lead wolf returned to the others and then headed toward the south arm of the lake, which it reached minutes before the rest. It raced down the ice to a point downwind of the moose, where it sat and waited for the others. When they appeared, the leader ran to meet them. All stood, nose to nose, and wagged tails for a few moments. But instead of pursuing the moose, they trotted to the middle of the lake and curled up on the ice. Sometime later the other five wolves followed the trail made by the first group and met them on Lake Richie, where they all lay down and rested.

After an hour or so, the wolves stirred and followed an old trail to Intermediate Lake, with tending and mating along the way. Again, whenever the pack moved, there was one wolf that seemed to lead them. Although it broke trail through some deep snow, it still was always 25 to 100 yards ahead of the rest of the pack. When the pack was involved with mating activity, this wolf usually looked back and waited. At least a couple of times we noticed a large and a small wolf traveling close together but staying with the pack.

As the pack traveled, the lone wolf followed, usually staying a couple of hundred yards behind. Sometimes the lone wolf would be able to join one or two of the last wolves in the pack, which themselves seemed to rest a lot and showed little interest in mating activity. The lone wolf was loitering with the other two, apparently upwind of the rest of the pack, when two wolves from the main pack suddenly began pursuing it at top speed, running right by the two other wolves and cornering the lone wolf near a snow-bank. They immediately attacked, but the lone wolf fought back. Within seconds, the fighting stopped, and the two aggressors rejoined the excited pack, with much tail wagging and nose-to-nose contact. The lone wolf followed slowly, and again the two wolves attacked it momentarily. This happened a third time, after which the lone wolf did not attempt to follow the attackers.

I wondered about the two wolves that seemed to tolerate the lone wolf. Were they immature wolves that didn't care yet about mating or competition? Or were they old and senescent, without much fight left in them?

Later, about a half hour before sunset, ten of the wolves were traveling in the woods, not far from the shore of Siskiwit Lake, while the remainder straggled about a mile behind. Suddenly, the leading group stopped and sniffed the air. Across the wind, three adult moose browsed about three-eighths of a mile away. The wolves headed away from the lake single file to an old beaver meadow. They traveled downwind a few hundred yards, veered, and continued for 250 yards until they were directly downwind of the moose.

Then they bolted toward the browsing moose. With the wolves just 25 yards away, two of the moose ran. The wolves did, too, but stopped when they saw the third moose, which had remained behind. They swerved to surround it. Finally the moose ran, the wolves following. Soon five or six animals were biting at the moose's hind legs, back, and flanks. The moose dragged the wolves until it fell, and the wolves swarmed over it. The moose jumped back up but fell again. It jumped back to its feet and ran through scrubby brush toward a thicket of spruce and aspen while the wolves continued their attack.

It was hard to keep everything in view, but Don was very good at circling the plane so I could continue to watch the chase unfold below us.

Then something happened that I never expected, had never read about, and amazes me still. One of the wolves sunk its teeth into the rubbery end of the moose's nose, effectively anchoring the moose as the other wolves attacked its rump and rear legs.

The moose finally shook the nose wolf and reached the thicket, where the wolves had no room to maneuver. The wolves retreated and lay down nearby as the moose stood among the trees, bleeding from the throat. Two or three continued to harass the moose without actually biting it, and the moose retaliated by kicking with its hind feet. Whenever the animal faced the wolves, they scattered. Despite the bleeding from its throat, the moose appeared strong and belligerent.

At 6:30, as the sun set behind the hills of Minnesota's North Shore, we climbed above the fray below and reluctantly headed west toward camp. I so hated to leave. Don, too. What would happen while we were gone? Would the wolves attack again? Would the wounded moose just die, and then the wolves start feeding? Chances are we would never know the final details. What if it snowed and we couldn't get back to the scene for days? Still, we both knew the deadly peril of overstaying our watch. No matter how good the pilot, if it's too dark to judge exactly how far off the ice you are when trying to ease your skis down, you're probably doomed.

I was thrilled, of course, at what we had managed to watch. I had just witnessed one of the most interesting and exciting events a young wolf researcher could hope to see. Hardly anyone—any scientist, at any rate—had written about what we had just seen.

To our great relief, the weather cooperated. The next morning we immediately flew to where the wolves had cornered the moose the night before. We expected to see a dead moose, half devoured, the snow bloodied around the kill site. Or at the very least, the wolves and moose still locked in a standoff. But when we arrived, the wolves had vanished. The moose lay within 25 feet of where it had made its stand. After we flew several low passes, it proved to be very much alive. It rose to its feet and walked stiffly, favoring its left front leg. It was no longer bleeding and seemed in good shape.

Fast forward more than two weeks: We spotted the large pack feeding on a carcass a quarter mile from where we had last seen the gimpy moose. Had it died from its wounds to be found later by the wolves? Or had they returned some days later to take advantage of its weakened condition and finish it off? We didn't know. We couldn't even be sure it was the same moose.

Even now, after nearly a month of watching wolves, I was still unsure how tolerant they were of our presence. As we flew above them, they seemed to continue about their business of traveling, hunting, and mating. But I was curious how fearful—or bold—they might be when a human appeared at their level.

I soon got a chance to find out. We found the large pack of fifteen resting on the ice of Lake Desor near the southwestern end of an island. As we circled overhead, they arose and moved to the western end of the lake. As we settled onto the lake, about a quarter mile from the wolves, they seemed scarcely concerned at all. Then the plane backfired. The wolves jumped up, scrambled a few feet, and then lay down again.

I decided to get out of the plane and walk toward them to see if they would spook. Don did not like the idea. What if they didn't run? What if they decided to come over and check me out? The wind was blowing from us to the wolves, so surely they could smell me as well as see me. This was an excellent chance to learn more about what these wolves thought of a curious human. I had gone barely 100 yards when the wolves jumped up and ran to the east end of the lake. OK, I learned something. I trudged back to the plane. As we took off and circled, the wolves still appeared to be afraid and continued to look where the plane had landed.

Don and I continued to check on the wolves on Lake Desor throughout the day. Late in the afternoon, the wolves rousted themselves and seemed to be sniffing the wind. They soon began filing along the lakeshore. Unfortunately, we were low on gas, so we flew off to Mott Island to refuel. In preparation for our study, the park had stashed several drums of aviation fuel for us with a hand pump on the island dock. We landed and took turns pumping gas into the plane.

When we returned to the wolves less than an hour later, we saw they had surrounded a moose standing in a clump of aspen. Its head hung as though it had difficultly holding it up. Blood had soaked the snow around it, and it was bleeding profusely from its throat. Its lower left hind leg was bloody, and it leaned against a tree, keeping its right hind leg centered beneath its weight.

Most of the wolves stood several yards away, resting or playing or licking the bloody snow. But one wolf, its legs covered in blood, continually harassed the moose. It circled close and nipped at its quarry's injured leg.

Unfortunately, darkness was falling, so we had to scoot back to the base. Poor weather the next two days kept us grounded. When Don and I returned to the site three days later, we found the bones of the moose scattered around the spot where we had last seen it. Only several days later did we get a chance to land nearby so I could snowshoe in to the kill. There lay the remains of an old bull. Judging by the wear to its teeth, I put its age at eight to fifteen years.

The incident revealed how wolves could effectively dispatch even a mature bull. That certainly squared with the prevailing narrative that a pack of wolves was an efficient killing organization. But, of course, I already knew that wolves killed a wide range of large animals, including mature moose. Perhaps more illuminating were the times we watched the pack break off an attack or never even take up an apparent opportunity to give chase.

In early March, Dave Stimson replaced Bob Linn, who flew back to the mainland. Don Murray had flown back, too, as a new pilot, Lee Schwartz, arrived to fly the Aeronca Champion. I hated to see Don leave. We had seen so many interesting things together, and sharing meals and most of our time together, we had become very good friends. I didn't even know when I would see him again, if ever.

Late that very afternoon, Lee and I headed out and found ten of the wolves traveling the shoreline near Rainbow Point, at the far southwestern end of the island. Suddenly, several pointed inland, noses in the air. They seemed to have detected five moose, the closest only 30 yards away. As the wolves watched with wagging tails, all the moose ran into a nearby stand of dense spruce. Two wolves took several steps toward the moose but then stopped. Clearly the wolves were close enough to attack, but the leaders went no farther. The moose had been feeding in a clearing full of blowdowns and deep snow, which may have been why the wolves didn't follow. When the moose entered the spruces, still only 150 yards away, they stopped and looked toward the wolves, which had continued along the shore.

Not only did wolves sometimes inexplicably fail to give chase, but they also failed much of the time when they did decide to attack. In early March, Lee and I watched ten of the large pack traveling the shore of Isle Royale in Malone Bay. The leading wolf started toward Malone Island, a few hundred yards away, but suddenly stopped and turned toward the main island. A few moments later it was chasing two moose 125 yards inland. The moose separated, and the wolf chased one 125 yards farther, coming within 30 yards of it, but as the moose entered some spruces, the wolf stopped and returned to the pack on shore. All wolves assembled, wagged tails, and ran to Malone Island, a small dog-bone-shaped piece of land.

The ten wolves filed across Malone, directly toward a cow and calf lying on the opposite shore. From our vantage, it seemed the wolves had already scented them. As the pack closed to within 100 yards, the cow rose and ran to the calf a few yards away. The wolves surrounded the moose but held off their attack. Slowly the moose moved to thicker cover. The cow stayed close to the calf, protecting its rear. Several times she feinted toward the wolves, causing them to scatter. The wolves lunged at the moose for several minutes but never really made contact. Then they retreated to the ice, where they gathered up, wagged tails, and lay down. Darkness forced us back to camp. But when we returned the next morning, tracks showed the wolves had made another try. They had chased the moose onto the ice, where a large area packed with wolf tracks suggested the moose had stood off the wolves for some time. We saw no signs of blood. Tracks showed the moose had left the island from the north shore, and the wolves from the west end.

It was weird. Several times now the mighty pack of killers had begged off on attacking what seemed to me to be worthy prey. What was going on? Clearly they were hungry, for I had watched them long enough to know that once they cleaned up a kill and left, that meant they were ready for another. It took them two to three days to tear apart a moose, feed several times (up to 22 pounds at a time, I was to learn many decades later), chew the bones, and then head on to find another meal. So why pass up so many possibilities?

Through the rest of the winter, as we circled above the large pack, we often noticed the lone wolf tagging behind. Clearly it was accepted by a few members of the big pack but scarcely tolerated or aggressively repelled by others. It was apparent that despite its frequent appearance, it never became a part of the pack.

When we first spotted the outcast wolf, we also noticed two animals—young, old, we couldn't tell from the plane—that seemed perfectly willing to associate with it. On another occasion, when most of the animals were resting, two wolves backtracked around a point about 25 yards to meet the lone wolf. From our altitude, of course, we had no way of telling if these were the same two wolves that had earlier befriended the lone wolf. But they sniffed the cowering individual a few seconds and accepted it. The three then moved back around the point toward the rest of the pack.

When the lone wolf saw the other wolves, it ran about 25 yards, lay down, and remained there.

On another occasion, ten members of the large pack had begun traveling while five wolves rested. These five included three lanky, lighter-colored animals that we suspected might be juveniles. The lone wolf joined these five. When they finally rose to travel, the lone wolf joined them and melded right in. There was no apparent friction between them.

A similar situation occurred a few days later. Six wolves, including the lanky individuals, were several miles behind the rest of the pack. The lone wolf joined these and accompanied them, without any apparent fear, to the rest of the animals. When they approached the main pack, the wolves engaged in their usual sniffing, tail wagging, and other greetings. But the lone wolf ran off and stayed away from the others.

Even though the lone wolf was ostracized by much of the pack—or at least by key members of the pack—it still tried to join in the pack's activities. Once we even saw it follow the group in pursuit of a strange wolf. All sixteen animals chased the alien for half a mile.

The lone wolf was interesting because it was an anomaly. I believed that no wolf would choose to be isolated, and this wolf gave every indication of wanting to join the pack. For convenience, we called the lone wolf Homer, though for the life of me I can't recall how we chose that name.

Where had Homer come from? Had he once been part of the pack but had been driven out? Was he always of low status and had been kept at a distance? Might he have traveled alone over the ice to Isle Royale and now desperately longed to join a pack to more successfully hunt and better survive? We simply didn't know.

I was aware that in the early 1950s, four wolves had been introduced to Isle Royale from the Detroit Zoo. The reintroduction plan, a long time in the making, was a bit of an anachronism by the time it was carried out, since wild wolves had already gained a presence on the island. Shortly after the four wolves were released, the Park Service decided that the zoo wolves, habituated to humans and not well suited for life in the wild, should be removed. Two were shot and killed. A third was trapped and removed. The fourth was never captured. Dubbed "Big Jim," he had been raised by Lee Smits, outdoor editor of the *Detroit Times* and wolf advocate. Big Jim, reportedly, had developed into a fine retriever of ducks. At any rate, Jim, who weighed 90 pounds when only eight months old, vanished into the Isle Royale forest. I wondered if now, six years after his release and escape, Jim was the mysterious lone wolf trying to find his pack.

James Cole, the biologist who used a plane to try to count wolves and moose on Isle Royale, had similar thoughts. Two winters before I started my work, Cole and Jack Burgess were flying the island when they landed on Siskiwit Lake and broke their landing gear. Unable to take off, they began snowshoeing back toward camp. For six hours, as they crossed Siskiwit Bay, a lone wolf followed a quarter mile behind, stopping and lying down whenever they rested. Wrote Cole, it acted like a "friendly but cautious dog."

Dr. Allen's chief directive to me was, "Count wolves, count moose. Learn all you can." Easier said than done, of course. But after several weeks of flying, I was getting a pretty good picture of the island's wolf population. There appeared to be a main pack of fifteen animals, which we could find and identify many days, as well as the lone wannabe member of the pack that tagged along behind and was on good terms with a few of the members of the big pack. And then there appeared to be two much smaller packs, of two and three wolves each. So, twenty-one wolves. I felt that was quite an accurate number.

But counting moose was a much different matter. For one thing, there were a lot more of them, which is a necessary fact of any prey–predator system. And unlike the wolves, which hung in discrete packs, the moose scattered across the island as individuals or cow–calf pairs. Finally, despite their size and dark color, they could be rather tough to see from the air if they sheltered in conifers or stood in shadows.

Early attempts to census Isle Royale moose were merely edu-cated guesswork. Larry Krefting and Shaler Aldous attempted the first aerial survey I was familiar with, flying a portion of the island in 1945 in a Waco five-seater biplane with a massive radial engine. Krefting and Aldous covered about 30 percent of the island, ex-trapolated to the entire area, and added a 20 percent fudge factor to come up with an estimate of 510 moose. James Cole flew the entire island in a more modern Piper Cub and, by flying low to flush out hiding moose, tried to count every last animal. He spotted 242 moose, saw the tracks of another 48, and put his estimate at 300.

Keep in mind, of course, that moose numbers vary year to year with the condition of the browse and severity of winter. And no one had devised a way at the time to double-check the survey results—to "ground truth" the census. So any number was taken somewhat on faith.

We decided to try Cole's method. So beginning in early March, after we had located the wolves and watched them for a while, Lee Schwartz and I began flying our moose census, the most grueling and tedious task we had in the air.

Like Cole, we divided the island into convenient-sized plots with natural boundaries. With the Champ traveling 70 to 80 miles an hour at about 400 feet, we flew strips parallel to the length of the island. The strips varied in width with changes in terrain and forest cover but averaged about an eighth of a mile wide. The strips were only a few miles long, so we could remember the location of the moose we spotted and avoid counting animals twice. It took days to cover the entire island, and it was certainly possible that moose wandered out of one strip and into another to get counted again. But I assumed these movements were random—one moose counted twice and another not counted at all—and it would all even out in the end. Each time we saw a moose, even in an open area, we circled and buzzed it to flush out any animals hiding near-by.

This continued for nearly a week: 28 moose one day, 53 moose another. We landed to refuel frequently. And we were often grounded by snow and high winds. Because our fuel supply was low, I managed to count only about two-thirds of the island, spotting 176 moose. We spotted another 18 on the remaining part of the island incidental to our wolf tracking. So there appeared to be at least 194 moose on the island, a number I assumed was low. Even so, that would represent a greater density of moose than just about anywhere in North America.

When our moose count was finished, it was time to fly back to Eveleth and take the Greyhound bus home to Purdue. The long trip back gave me a lot of time to think about what we had accomplished.

First of all, the past six weeks were a validation of what we had set out to do. We had proved it was feasible to count wolves and moose from the air. More than that, we could track them from the air, calculate the distances the wolves traveled, follow their routes, and watch their behavior, including their attempts to hunt.

Furthermore, this was all brand-new. Murie had been able to watch wolves around their den but couldn't have hoped to follow them for miles and miles and watch them hunt. Cole and Stenlund had pioneered the use of light airplanes to spot wolves, but neither had invested the time and effort required to actually follow the animals and study their habits for days on end. What we had seen was groundbreaking. Being green at this, I was excited to be learning new things, knowing that I was gathering suitable information for my dissertation. I was far from drawing any conclusions, but I was amassing lots of good data—maps of wolf travels and moose distribution, sightings about mating behavior, observations on how wolves hunted, information about the health and age of moose that fell prey to wolves. After just a single season, the information promised to be voluminous and new—not just to me and Dr. Allen but to science.

I had also come away with some impressions that I felt might be backed up by more observations and become truly insightful. I was impressed by the moose that had managed to stand off the wolves overnight and, when we had returned the next morning, had stood up and walked away. I remembered the number of times we had watched wolves begin to pursue a moose but then decide to call off the attack. Or wolves that had given their all in pursuing a moose but had failed to bring it down. A pack of fifteen wolves— most people, and many biologists, would assume they could kill what they wanted, when they wanted. But our observations were beginning to show otherwise.

It was a lot to think about on my way home. It made me confident about the future. I had a good shot at completing my PhD. It was satisfying for a young guy just out of college. Like the bear project I had worked on in the Adirondacks, this was science. The big difference was that the Isle Royale project was *my* science.

The author, a young graduate student in 1958, surveys Sargent Lake from the heights of the Greenstone Ridge near the northeast end of Isle Royale.

A storm approaches as *Ranger II* docks at Rock Harbor, at the island's rugged northeastern tip, in 1958. Later that year, the new *Ranger III* entered service.

The wildlife survey crew loads the Cessna for the initial trip to Isle Royale in February 1959.

Don Murray, a pilot from northern Minnesota, flew the Aeronca Champion for much of the winter 1959 survey and continued as the primary pilot for the wolf–moose study project for twenty years. Photograph by Durward Allen.

The author's plane approaches Beaver Island and Washington
Harbor, where the crew will settle in for the winter 1959 field season.

Wolves travel on frozen Lake Superior near the Siskiwit Bay dock.

A curious wolf approaches the fish house at Fisherman's Home, where the author hides with his camera in winter 1959.

A startled wolf flees as the author steps out from the cover of a fish house to snap this photograph in winter 1959.

The author cuts away the mandible of a moose killed by wolves. By examining the mandible and teeth, researchers were able to estimate age and health of the animal.

David and Betty Ann Mech (now Betty Ann Addison) spent the summer of 1959 in a guest cabin owned by commercial fishermen Ed and Ingeborg Holte on Wright Island off the southeastern shore of the island.

A bull moose grazes. Its developing antlers are still covered in the nourishing skin called "velvet," which will slough off as the antlers harden in fall.

Betty Ann washes laundry in the "clothes grinder" at the summer
home on Wright Island.

Left: Sharon Mech bathes in a bucket on the dock of the Wright
Island summer home. Right: Even in summer, the author bundles
up in a wool mackinaw jacket and winter cap. Photographs by Betty
Ann Addison.

The *Wolf,* a 16-foot runabout, allowed the author to quickly travel from one end of the island to the other—when inclement weather and mechanical breakdowns didn't intervene.

The sun rises in February 1960 at the Windigo ranger station, where the research crew has set up its base.

Pilot Don Murray tends to the Aeronca Champion on Hatchet Lake in the interior of Isle Royale, where he and the author set down on the ice.

Researchers shuttled gear across the ice to their cabin aboard a Studebaker Weasel, a military-surplus tracked vehicle.

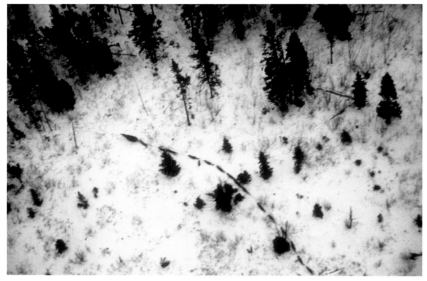

A large pack of wolves pursues a moose through spruce and brush near Grace Creek, not far from the Windigo ranger station.

Standing its ground, an adult moose holds the large pack of wolves at bay. After five minutes, the wolves gave up and left.

A large pack of wolves succeeds in killing an adult moose.
Photograph retouched by Betty Ann Addison.

The author poses in 1960 with nineteen of the many moose jawbones
he collected around the island to help determine age and health of the
animals at the time of death.

Red foxes were abundant on Isle Royale during the time of the author's survey and frequently hung around the winter camp.

Like other island residents, the author chopped ice from the lake in spring and stored it in sawdust before cutting it into smaller pieces for home iceboxes.

The Bangsund house on Rock Harbor, where the Mech family began staying in the summer of 1960. Photograph by Betty Ann Addison.

Commercial fishermen Pete and Laura Edisen were longtime summer residents of Isle Royale and became good friends with the Mech family.

A full-grown bull moose lumbers to shore from Ojibway Lake.

The author (left) and Durward Allen examine a moose jaw.

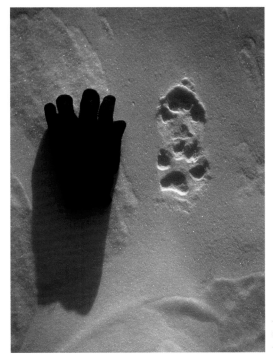

A glove provides scale for a pair of wolf tracks in the snow.

Members of Isle Royale's large wolf pack show little concern for the author as he takes photographs.

ISLAND LIFE

SUMMER 1959

❧

For all the reasons Isle Royale made a perfect laboratory for
the study of moose and wolves, it also made an idyllic home
for two young newlyweds who washed up on its shores. Betty
Ann and I lived during this second summer in a tiny log cabin on a protected bay of Wright Island, along the main island's
southern shore. Like the wolves and moose, we were isolated,
limited in our comings and goings. Like the animal community,
our lives were stripped down, simpler than they were back on
the mainland.

We had arrived from Houghton in early May aboard the
Ranger III, the Park Service's brand-new 165-foot boat, put into
service the fall before to ferry cargo and passengers to the island.
Two days later, park ranger Dave Stimson shuttled Betty Ann
and me down the long shoreline toward our new home a couple
hours away from park headquarters by boat. But in a morning
that would foreshadow the difficulties I would have with small
boats, the outboard crapped out near Chippewa Harbor and
would run only at low speeds. Dave couldn't fix the problem,
so we puttered back toward Mott Island as Dave radioed park
headquarters. The *Tonawanda* met us just as our motor quit entirely. The deck hands threw us a line and towed us back to
headquarters.

As we ate lunch at the mess hall, Frank Taddeucci repaired
the outboard, and by 4:30 we arrived at our new home.

Wright Island had been fished most recently by the Johnson
and Holte families. The family summer home, situated up the
hill, had been built in the 1920s by fisherman Steve Johnson for

his wife and father, when they began commercial fishing the area. Steve Johnson's daughter, Ingeborg, and her husband, Ed Holte, eventually took over the property and raised their daughter, Karen, there during the summers, when they netted the nearby lake.

Our abode, unfortunately, was a small log cabin, even more rustic than the Holte summer home. My wife called it a "hovel" and said the row of long windows reminded her of a chicken coop. It looked out over the dock and fish house situated on Hopkins Harbor. There were also a net house and guesthouse. A tiny privy stood out back. By the time we arrived, the island's lake-trout fishery was in steep decline, and the Holtes weren't fishing through the summer, though Karen would visit the island for a couple of days in early July.

Ranger Roy Stamey stopped by to help us get the propane refrigerator running. We moved our groceries onto the shelves and into the fridge. We moved furniture around to better suit our needs and stored a bed and daybed in the larger house to make more room in our little cabin. Roy also helped me fire up the "clothes grinder," a washing machine powered by a single-cylinder gasoline engine. Periodically I "mowed" the lawn around our cabin—a rank thicket of fireweed, asters, and other weedy growth—with a scythe.

At any given time during the summer, a couple of hundred visitors would be hiking and camping on the main island. But they rarely made it to Wright Island. We really were quite alone. Our cabin sat nearly equidistant from the park headquarters to the northeast, and Washington Harbor to the southwest. Malone Bay ranger station sat only 2 miles away, manned by Ray Wiinamaki, but was hidden behind several smaller islands.

Stamey and other rangers would drop in from time to time. The *Voyageur*, a tug-sized boat, from Grand Portage, Minnesota, came by twice a week. It made its way around the island picking up fish from the few remaining commercial fishermen and delivering their supplies as well as carrying visitors.

For Betty Ann and me, the *Voyageur* was our main lifeline to the rest of the world, bringing groceries, mail, newspapers, and gawking passengers. Imagine being a well-to-do tourist riding around a large island national park, briefly seeing a few elderly Scandinavian fishermen living in weather-beaten cabins and then coming upon a young couple living in a shack on an isolated island many miles from civilization. We always wondered what they might have thought we were up to.

Otherwise, we would motor 5 miles down the shore to Hay Bay to pick up herring for breakfast from John Skadberg, one of the few remaining commercial fisherman. One evening a sailboat with a crew of eight landed at our dock for supper. That ranked as a major social occasion.

Sun rose over our snug harbor. We would watch the beavers swimming in front of our dock. Occasionally one would waddle ashore. We felt like castaways on our little island. But Betty Ann was game for this. She had put her academic career on hold and had worked in a Purdue lab while I was in graduate school. She enjoyed this time away from the mainland as much as I did.

During fall when I took courses, Betty Ann was a lab technician for an eminent biophysicist. She ran precisely timed, milkshake-like devices full of mixtures of bacteriophages, types of viruses that were drilling their way into bacteria. Precise timing of the brew would then allow only certain parts of their DNA into the bacteria, thus allowing her supervisor to learn the traits of that part of the viral DNA by observing the change in the bacteria.

Her supervisor was Professor Seymour Benzer, who we later learned became the father of neurogenetics. He was furthering early findings that had won a Nobel Prize. Only long after I had completed my PhD work and begun collaborating with geneticists on wolf DNA did I fully appreciate what Betty Ann had been helping with in the lab.

Betty Ann was about six months pregnant when we arrived on Isle Royale, which meant we planned to return home just days before her due date. There was no resident physician on the island, but several doctors visited in summer as tourists, many on their own yachts. The Park Service rangers, whom we knew well, always made sure to notify us when a visiting physician would be in our area so I could arrange for them to examine Betty Ann.

On one occasion, a doctor's yacht was moored at a dock at Malone Bay, a few miles from our home on Wright Island. I zoomed over there with my boat and asked the doctor if he could examine my wife. He said that at this stage in the pregnancy all he needed to see was a urine specimen, which he would analyze back home in Houghton. So I headed back to Wright Island and obtained a urine specimen. Sensing the possibility for a practical joke, I took some orange juice concentrate and in a gallon jug made a very diluted batch of juice. I assumed the doctor's highbrow friends would regard anyone living primitively in a shack as I was, not as a budding scientist but as some kind of yokel.

Heading back to the yacht, I kept the real urine specimen in my pocket and the gallon jug next to me on the seat where it could be easily seen as I pulled up to the dock next to the yacht. As the doctor and his partying friends gathered to help me moor the boat, I picked up the jug and asked, "Is this enough, doc?" As the doctor's buddies guffawed at my apparent ignorance, I slipped the doctor the real urine specimen, untied the boat, and left.

Later in the summer, I met a Detroit obstetrician who was vacationing at Rock Harbor Lodge. He and his wife agreed to come all the way down to Wright Island to examine Betty Ann. After the impromptu exam, the doctor said Betty Ann was doing well and that the baby would probably be born late. They stayed the night, and the next day we all motored to the Malone Bay dock and hiked along Siskiwit Lake.

His examination was a relief to us, though in retrospect I'm not at all sure how he would know such a thing. I can't say we felt much trepidation about our rather tight schedule on the island, but we were young and perhaps naively confident.

My second summer on Isle Royale found me in a funny position. I already knew the island from my first summer hiking the trails. And it was clear from last winter's work that the meat of my research came from flying over moose and wolves when the leaves were down. So, what to do? Collect wolf scats to see what the animals were eating during the summer. Gather up any gnawed moose bones I could, especially from obvious wolf kills, to determine the health and age of the moose that fell to predation. Compared with the rigors of logging long hours in a flimsy plane in cold, windy conditions, my summer seemed like a different project—more of a working vacation.

Well, not entirely. One of my first efforts was to examine "Dead Moose 14," a cow that had been found a couple of weeks earlier. The moose had been pregnant, and when I cut it open, I discovered two fetuses, which I dutifully weighed, measured, and photographed. I cleaned the meat off the mother's mandible to save it for aging, and preserved and labeled the specimens.

Otherwise, I hiked the trails like the past summer. Roy Stamey and I got word of a dead moose near Siskiwit Lake. We couldn't find it but did run across another old kill. We visited another that Roy had found the year before, but we couldn't find the mandible. And so it went, rather routinely. Most of the time, that is.

It was bound to happen sooner or later. And it happened early this summer. After just two weeks on the island, my old back injury came back to haunt me. On May 28, I had hiked 24 miles collecting scats, and the next day I was a bit sore. Thus I decided to skip one side trail and just head the 10 miles back to my pickup point. On the way back, I managed to reinjure my back and had to hobble to Crow Point, where I had arranged to meet Betty Ann and Roy, who took us back to our cabin. I had to rest several days, working at the cabin and puttering with the boat, before I was fit again for any long-distance hiking.

Betty Ann accompanied me on a lot of these trips, collecting and identifying plants, painting landscape scenes and wildflowers, and making plaster casts of wolf tracks. She had the idea of making ceramic wolf-track ashtrays from the casts. Later in the summer, she got word that she could begin selling her paintings and the ashtrays at the Rock Harbor Lodge.

Whenever I traveled, I would often stand on overlooks, surveying the surrounding forest with my binoculars in hopes of finding a wolf den. I wanted to watch the wolves as they raised pups, as Murie had done in Alaska. But I realized that in the dense boreal forest of Isle Royale, where visibility was limited and it was nearly impossible to track wolves over long distances by sight, finding a den would be largely a matter of luck. I really didn't know where to begin.

Roy and I were exploring the eastern end of Siskiwit Lake one day in mid-May in our quest for dead moose. Scanning the ridges beyond the north shore of the lake, I spotted what appeared to be freshly excavated dirt. It was too windy that day to cross Siskiwit in our outboard motorboat, but I noted the location. Several days later, Roy and I crossed the lake and found the site. It was indeed a large den. The den entrance was big—more than 2 feet wide and 16 inches high. The tunnel leading inside was a foot wide. We found an old moose calf mandible nearby. But I was disappointed to see that the tracks we found in the fresh dirt were those of foxes, not wolves.

Most everywhere I hiked, I found wolf tracks. The previous summer, having never seen a wild wolf, I was excited to find the tracks. But now I had seen the wolves. I knew about how many lived on the island, how they hunted, and where they traveled. During summer I found the lines of tracks as singles, or in pairs, suggesting that during this time of year the wolves probably set out from the den (wherever it was) alone or in pairs to hunt rather than traveling as a pack. Otherwise, the tracks, of themselves, signified nothing I didn't already know. In fact, they became a source of frustration. What could I do to make these tracks tell me something? I continued to put out meat scraps to attract foraging wolves, in hopes the direction of the tracks might somehow lead me to a den. I even considered tying a long fishing line to a scrap to get an idea in which direction the wolf was taking the meat. But for now, tracks were just tracks, evidence that a wolf had been here. Something I already knew.

But scats, they were something else. As smelly and messy as it was to gather dozens of scats along the trail, they provided real information. They were a simple objective record of what a wolf had been eating. And I could gather them by the hundreds—enough to actually provide data about the wolf population as a whole.

Yet, as with the tracks, I wanted them to tell me more. Specifically, it would be handy to know how old some of these scats really were. A day? A month? I didn't really know unless they were fresh out of the wolf, still moist.

So I collected a fresh wolf turd. It contained a lot of hair and also a lot of soft, wet, shiny organic stuff. I brought it back to the cabin, but instead of bagging it and drying it, I set it outside and let it sit.

I tried the same with two more scats to see if they would age differently. Referring to them as the "experimental scats," I made notes of how their appearance changed over time, the better to judge the age of the scats I found along the trails. After five weeks, the first had shrunk and become hard, black, and dry. And that was it. The next month and the month after that, it looked pretty much the same. After a month, the second scat appeared partly white, partly light gray. Nearly a month later it had started to fall apart but otherwise looked much the same. The third scat began turning white and breaking into small pieces after a month. A month and a half later, the hair and bone were obvious, and part was covered with chalky material.

So now you know next time you find a wolf scat.

When I had flown above the island the previous winter, I had often found that wolves would return to their kills, sometimes for days on end. That raised the question, at what point do predators become scavengers as well? A moose carcass I found with Lee Smits, the newspaper editor who had raised "Big Jim" and was visiting me in late June, convinced me that wolves were more than willing to survive by scavenging.

Planning to hike to Mount Ojibway fire tower, Lee and I landed my boat at Daisy Farm Campground, where we met two hikers. They said they had just crossed the island from McCargoe Cove and had discovered a "wolf-killed bull moose" near the Chickenbone Lake portage trail. So Lee and I changed plans, returned to the boat, gassed it up at Mott Island, motored 27 miles around to the north shore of the island, docked at McCargoe Cove, and hiked in to Chickenbone Lake. We found the dead moose, its chest and head lying in the water. Its antlers were fully formed, without the velvet that covers them in summer, evidence that the animal had died the past fall. (Moose shed their antlers each winter.) Its guts and leg bones, covered with putrid meat, lay in the matted-down grass on shore. I grabbed a stick and poked at what flesh remained on the bones in the lake. It quivered like jelly. We found wolf tracks nearby. I collected the mandible from the skull, took photos, and we left. A couple of weeks later I returned to the carcass with park naturalist Bob Janke. The wolves had dragged the head and chest onto shore. The flesh was almost entirely eaten. The wolves had eaten all that putrid meat.

One day I carried my backpack from the Siskiwit Bay CCC camp up the Island Mine Trail, collecting wolf scats as I walked. Reaching the Greenstone Ridge Trail, I found tracks of at least one and possibly two wolves headed both east and west. I turned east and hiked toward Lake Desor. Reaching the west end of the lake, I scanned the ridges in all directions but saw neither a wolf nor evidence of a den.

I bent down to collect a fresh wolf scat on the trail to the lake. Straightening up, I thought I saw a flash of movement ahead of me. Up ahead I found fresh wolf tracks. I ran to the west end of the lake, looked around, and saw, standing about 200 yards away, a large wolf. It followed the trail I had just traveled westward for several hundred yards. Peering through binoculars for several minutes, I watched as it followed the trail through a swamp. Then, as the trail threaded through heavy timber, it disappeared.

That was the first wolf I ever saw as I hiked the trails in the summer.

A couple of weeks later I had taken my boat down to Windigo at the southwestern end of the island and begun collecting wolf scats along that end of the Greenstone Ridge Trail. Again, there were fresh tracks heading in both directions—and fresh scats, too. I heard what sounded like whimpering and whining several yards behind me. I turned and saw nothing on the trail, so I hid behind a tree and waited. Nothing. I continued up the trail. After another 2 miles, I suddenly came face-to-face with a large wolf. It stood on the trail 50 feet ahead, looking directly at me. We watched each other about—what?—ten seconds. Then it vanished. I can't even recall how or where. It was gone. I rushed to the spot where it had stood but could neither see it nor hear it.

That was the second time.

Just three weeks later, I was camped at the Lake Desor lean-to, a primitive log shelter with a roof and three sides. I cooked and ate breakfast and then walked down to the lake. On the beach, about 100 yards away, a single wolf was walking north, away from me. I watched for perhaps a half minute until it entered the swamp where the beach ended. I waited for more than an hour, but I did not see it again.

That was the third wolf.

It was thrilling to hike into the woods and see a wolf—no airplane, no noise, no modern contrivance. Just the wolf and I, both walking on foot to meet at a single place in the wilderness. But by now, actually seeing a wolf told me next to nothing. A sighting, like a track, was just an ephemeral sign that a wolf once had been here. But I loved seeing them anyway.

I liked observing other wildlife on the island, mostly for my own interest rather than any relevance to the moose–wolf dynamic. As I gathered wolf scats, I often also picked up the much smaller scats of foxes, the only other canine predator on the island and potentially a competitor with the wolf for smaller prey such as snowshoe hares. Along the way I also made note of other wildlife—eight blue-winged teal one time, a woodcock on a nest, peregrine falcons on occasion. And tracks of what appeared to be an otter, an animal that by all rights should have been here on the island since they can swim forever, but not positively identified here in an awfully long time. Along Siskiwit Lake, I found a fresh, partly eaten male red-breasted merganser and saw a mink—the animal that most likely captured it—diving into a small stream.

I spotted a pair of bald eagles on their nest, about 60 feet up a large aspen that had recently been struck by lightning. (The bald eagle population across the continental United States had plummeted during the past decade, even in the wilderness of the Great Lakes region, for reasons we did not yet understand.) I watched for about twenty minutes as the irate birds circled above the nest. I took note of snowshoe hares I spotted, realizing that they could be a source of food for wolves, though probably not an important one.

And I was always surprised by the sharp-tailed grouse I saw. I once spotted five within just a couple of miles of the Siskiwit Bay CCC camp. I remained surprised that a bird more typical of grass and brush lands far to the west and south would be common on Isle Royale when ruffed grouse and spruce grouse, true denizens of the north woods, were nowhere to be found.

It's hardly surprising that I would see beavers as I hiked past lakes and ponds and cruised the shoreline in my boat. They are ubiquitous in the Great Lakes region. And since few mammals swim better than a beaver, it's natural that they would have found their way to Isle Royale. From what I could tell in talking to long-time island residents, beaver numbers had waxed and waned. They were apparently more common before wolves found their way to the island, and some fishermen suspected wolves were the cause of the beavers' decline.

Of course wolves would eat beavers. They're a big piece of meat—an adult can weigh 50 pounds—and are relatively defenseless on land. So I wasn't surprised that in the seventy wolf scats I had collected, dried, and analyzed during my first summer, beaver bones or fur made up 17 percent of the food items I found.

Most of the time, beavers are safe from wolves, swimming in relatively deep water or tucked into their lodges in winter, eating from their submerged cache of aspen and other species. But they do have to come to land at times to gnaw down trees, and that's when, I suspected, they fell prey to wolves.

Because of their relationship to wolves, and the possibility that wolves may be controlling their numbers by eating them, I was interested in learning more about the island's beaver population. As I traveled around the island, I often made notes about any beaver activity I saw. Sometimes I found fresh beaver cuttings around ponds and lakes. But other times I found large dams that had no fresh cuttings. It appeared the original residents had been killed or moved on. As dams were abandoned, the ponds they created would gradually drain and turn into grassy meadows.

Once Bob Janke and I hiked along the inlet stream to the eastern end of Chickenbone Lake. It looked like a perfect spot for beavers. We found several old dams, some still holding water. Birch and aspen were abundant to within 100 feet of water's edge. But we found no fresh sign. Had wolves killed the colony? Had the beavers moved on to better food supplies? Or had they died from a contagious disease that had killed beavers and muskrats in much of the Great Lakes a few years earlier? It remained a mystery.

Having Bob Linn's boat, the *Wolf*, made it possible to visit the entirety of Isle Royale during the summer, including the many smaller islands that lie along its shore. But that versatility came with a price. It seemed I was forever puttering with the boat to keep it seaworthy and running smoothly.

It was a handsome little runabout, 16 feet long, of wooden lapstrake construction, with a windshield. It was powered by a 35-horsepower outboard, as I recall. We timed it once at 26 miles per hour, wide open. As I mentioned, I stowed a spare 10-horse outboard on the floor to get back to shore in case the main motor sputtered out.

And in those days, that was always possible. Outboards were finicky and troublesome. Because the outboards were two-cycle engines, I had to mix oil with each new batch of gas, which produced a cloud of smoke upon starting. Often the motor would run rough, misfire, or refuse to run at all. So I became expert at pulling the spark plugs, cleaning, and gapping them—sometimes doing it a couple of times a day. I also had to adjust the boat's steering linkage, fix things like the speedometer pressure tube, and do general maintenance like pumping out water from the bilge and cleaning the boat. And if all else failed, I would limp into park headquarters and ask Frank Taddeucci to fix whatever needed fixing.

A boat equaled freedom to travel, but wind and fog often changed that equation. Sometimes I couldn't leave our little island. Other times, I couldn't get back to it.

Situated on a little hook of land that faced a small harbor, our cabin was safe from the strongest wind. Even in a hellish storm, gentle waves merely lapped at our dock. But if I rounded the point less than a quarter mile away, the waves from a strong southerly blow could send me scurrying back to shelter. Betty Ann and I had barely set up housekeeping when winds of 44 miles an hour, with gusts to 60 miles an hour, pinned us to the island for the entire day. Around the point the wind lashed the shore with steep waves, making any travel along the shore in a small boat impossible.

I learned that patience was the most important ingredient of safety. That meant letting the weather decide whether to head out on a particular day—and feeling no regrets for staying home and searching Wright Island for old moose remains. Often, it meant that rather than heading home after a day in the field, I simply had to hole up somewhere and let the weather pass.

Once, starting out for Windigo, 30 miles away at the far southwestern end of the island, I chickened out as waves from the south built up in Siskiwit Bay and a storm appeared to be on its way from the northwest. So I landed at the protected cove at Fisherman's Home and visited Sam and Elaine Rude, a longtime fishing family, until late afternoon. The wind and waves subsided, as they often do late in the day. I took off and made good time to Windigo, arriving at dinnertime.

Wind is typically the least in midsummer—July and early August. But even then I was sometimes tripped up by the weather. One day Bob Janke and I were making the rounds of the northeastern shore. We stopped by Crystal Cove to talk to fisherfolk Milford and Myrtle Johnson. We continued on to Bob's place on Captain Kidd Island, where I spent the night. But the next morning, the sea was too rough for my small boat, so I helped Bob cut and stack wood. We collected plants, photographed a crow's nest, and in the afternoon picked up groceries the *Voyageur* had dropped at Emil Anderson's, taking some over to the Johnsons. By then the waves had begun to drop, so I set out to round the northeastern end of the island and head home, a 36-mile trip. Outside of Rock Harbor, as I was exposed again to the open lake, the swells were still huge, and I had to motor slowly, letting the boat nose over each swell and easing into the next trough to avoid taking water. By the time I reached the Malone Bay ranger station, it was ten o'clock and completely dark. I let the ranger know I was back and intended to head home. This travel at night was nerve-wracking. I picked my way through the reefs and islands to Wright Island and saw the kerosene lamp Betty Ann had lit to guide me in.

Sometimes fog, not wind, was the problem.

One day Bob Janke and I hiked the interior of the island and returned to the boat at McCargoe Cove, had a snack, and started back. I dropped Bob off at Captain Kidd Island on the northwestern shore and set off for park headquarters at Mott Island. But rounding Blake Point, I hit a bank of fog so thick I could barely see. Following the shore, I traveled down Rock Harbor and reached headquarters. I told Paul Richards, the park clerk, I planned to continue on. Finding my way out of Rock Harbor to the open lake, I hugged the shoreline, cutting through the narrow channels behind Schooner, Hat, Ross, and Malone Islands, until finally I reached the Malone Bay ranger station. The cabin on Wright Island was only 2 miles away, but I had had enough. I spent the night with Ray Wiinamaki.

In the morning, we set out to find three tourists who, Ray had learned by park radio, had set out in a motorboat from the tour boat *Bonnie Jean* and had never returned. We hugged the shore in the still-heavy fog, stopped by my cabin for a compass, and set out for John Skadberg's place to see if the men were there. Not finding them or John, we continued into Siskiwit Bay. We met John coming our way. He said he thought he had heard someone down the shore at the Siskiwit Bay CCC camp dock the previous night. We followed a compass bearing through the fog, arrived at the dock, and found the men, who had spent the night at the camp. The men followed us back to Malone Bay, and I returned home to Betty Ann.

Betty Ann understood the weather as well as I did, and when I didn't show up in the evening, she took it in stride, trusting that I had made a prudent decision to stay put.

Betty Ann was a good cook, and twice a week the *Voyageur* dropped off grocery orders from the mainland. Supplemented with regular fresh fish and occasional morel mushrooms from the trail, that all made for very good eating anytime I was home between trips. On top of that good eating, we were privileged to share a special treat enjoyed only by commercial fishermen who happened to catch whitefish. Whitefish are large, fatty denizens of the deep Lake Superior waters. Their livers, too, are fatty—totally different from any other liver I had eaten. Not only is their consistency much firmer than liver, but their taste is incredible. They are so good that the fishermen didn't sell them but rather kept them for themselves and as rare and highly valued gifts for certain friends. The whitefish livers were my finest culinary treat on Isle Royale. They are full of vitamin D, however, and if you ate too many, I was told, you would get headaches. I never was able to test that claim.

The fact that I now was married and brought my wife to live with me on the island for the summer opened up our social possibilities, compared to my previous year as a bachelor. Most every Sunday, weather permitting, Betty Ann and I motored some 25 miles into Rock Harbor to attend Father Murphy's Catholic mass. We usually stayed for breakfast, where we met many of the people visiting or working on the island during the summer. We often dropped in on fishing families, including Sam and Elaine Rude and their son, Mark. Traveling around the north shore of Isle Royale, whether alone or with Betty Ann, I usually stopped to chat with

Bob Janke of the Park Service on Captain Kidd Island, or ranger Terry Moore and his wife on Amygdaloid Island, or Emil Anderson in Belle Harbor. Dave Kangas, a young guy from the Upper Peninsula, led the trail crew each year. We became good friends, and I often dropped in on him at the Hatchet Lake patrol cabin, where his crew often bunked and ate. Occasionally a sailboat would dock at Wright Island, and often we were invited aboard for a visit. My parents and sister, who had traveled from central New York to visit us, stopped by the island during the summer, and we hiked in to Siskiwit Lake and looked for moose.

Bob Linn, the park naturalist, was the social catalyst on the island. He made a point of inviting Park Service employees as well as fishing families—and anyone else he could find, which included Betty Ann and me—to his house on Davidson Island. Bob had been on the island a long time, and by the time I showed up, he knew every fishing family who lived there.

An unanticipated phenomenon had happened over the past year, especially after my experience flying over the island in winter. I gradually had become a wolf authority.

Of course, as I have tried to make clear, the bar was set rather low. Few scientists had studied wolves, and none of them had been able to follow them around as I had. So after a year on Isle Royale, I had had experiences and insights few others could claim.

As I traveled around the island this summer, I often carried slides of the wolves and moose I had photographed the previous winter. I met a small group that included Lee Smits at Rock Harbor Lodge, the main tourist center of the island, and showed them my slides. At Windigo, a smaller tourist center at the opposite end of the island, I displayed my photos to Paul Brooks, a nature writer, environmentalist, and editor in chief at Houghton Mifflin. I presented slides to employees of the Rock Harbor Lodge and to members of the Isle Royale Natural History Association at their annual meeting.

OK, I wasn't exactly a movie star. But in a small way that was very important to a young graduate student, I was gaining the sort of stature in my field that the people in my still-small world recognized.

By mid-August, our sojourn on Isle Royale was coming to an end. There had been no big discoveries. Though I had managed to spot three wolves and nearly countless moose and other critters, there were precious few data to add to my thesis—except for what secrets might reside in the boxes of dried wolf scats I would carry with me back to Purdue. On August 17, we packed all our belongings and equipment and placed them aboard Bob Linn's larger boat, the *Loon*. And because the sea was rough, Betty Ann rode with Bob as well. I stayed behind and closed up our cabin before motoring to Davidson Island. It was a wet ride, the boat plunging into the waves coming from the southeast and hitting the boat roughly broadside. Betty Ann and I stayed with Bob and his wife, Fanny.

Soon, Betty Ann and I left the island aboard the *Ranger III*. From Houghton, she flew back to Purdue, while I followed in our car. Then, a week after we left the island, Betty Ann gave birth to our daughter, Sharon.

We weren't quite done with Isle Royale before the snow flew. In late October, Dr. Allen and I took the *Ranger III* through a heavy sea from Houghton. A skeleton Park Service crew were the only other folks remaining on the island, and they were closing up shop for the winter. We wanted to fly the island when the leaves were down for a "calf:cow" count. It wasn't part of the moose census that I had begun last winter but an attempt to better understand the productivity of the island's moose population.

We also wanted to try out a new gizmo Dr. Allen had brought along: a portable record player and a collection of records of wolves howling.

After supper the first night, Dr. Allen and I took the *Wolf* to Daisy Farm Campground, lugged the record player up the first ridge toward Ojibway Tower, and began to play the recordings. It was near sunset, but the wind continued to blow—not ideal conditions for the wolves to hear the records or for us to hear the wolves. Turning the record player off, we listened. About thirty seconds later, wolves howled far to the north. We played the records again, turning the device in several directions. Soon the wolves were howling plainly. We seemed to discern three wolves, and they sounded to be about a half mile away. At least one animal made high coyote-like yips. Over the next few days, we played the recordings several times more. Usually we heard nothing in response, though in one case, as we soon learned, Bob Linn heard our howls more than a mile away.

As sailors and fishermen know, Lake Superior turns stormy in the fall, and this fall was true to form. With the exception of a single calm, overcast day, the days were either windy or rainy. If the wind blew, we flew. If it rained, we didn't. In the wind, the plane bounced around as if it were a toy flown by an energetic three-year-old. Dr. Allen and I took turns looking for moose. In the turbulence, we both became violently airsick, filling the coffee can on each flight. Dr. Allen graduated from the can to filling his hat once, and when that was brim-full, he hurled down his jacket sleeve. On November 1, our last day of flying, the winds rose from 25 to 40 miles an hour. Jack Burgess, our very capable pilot for this trip, decided it had become too windy to fly. He flew me back to Mott Island, but the northwest wind was so strong he couldn't taxi on his pontoons. So he took off again and landed in a protected bay, where Dr. Allen and Ben Zerbey, the chief ranger, came by boat to take us back to park headquarters. Our fieldwork for the year was done.

FINDING PATTERNS

WINTER 1960

❧

It had been a long fall, replete with a full schedule of graduate classes among the Indiana cornfields, interrupted only by our short and not-so-sweet moose calf survey as well as numerous episodes of washing diapers. Because of my need to spend much of the year on Isle Royale, I had to double-up on classes. That meant taking twice the normal load per semester when I was at school in autumn. I was more than ready to resume my field-work.

My disappointment at having to leave my growing family now was offset by my anticipation of what the new winter of surveying the wolves and moose would bring. I headed out to Isle Royale in early February 1960 with more confidence than I had had a year earlier. Of course we would see wolves! I knew that. And I knew that what we had seen and learned the previous winter would put us on the wolves sooner and make our pursuit of them more effective. So I started the winter with greater expectations and a keener understanding of what we might see.

We would fly this winter with continued interest in wolf behavior: how they hunted, how far they traveled, how the big pack on the island interacted with the smaller groups of wolves on its periphery, the size and location of the packs' respective ranges.

But we also had an eye on the big ecological questions that this study provided us the opportunity to explore: Do the wolves kill indiscriminately, or do their limitations force them to focus their efforts on young, sick, weak, or simply unfortunate moose? Do they affect prey populations—in this case moose—only incidentally, or do they kill effectively enough to control the herd or even deplete it? What controls the wolf population itself? Is it starvation, or perhaps some innate controls on breeding and reproduction as well? And perhaps the biggest ecological and philosophical question of all: What is the role of the wolf in this wilderness area? These were the things we ultimately wanted to understand.

I would be flying again with Don Murray, who was showing a real personal commitment to the wolf project. With the experience of the previous winter behind us, I felt confident the two of us would be able to anticipate the movements of the pack and shadow it closely for days or weeks on end to get as clear a picture as possible of how far they traveled, how often they killed, and when they rested.

Bad weather forced us to cool our heels on the mainland in Eveleth for an entire day. The following day around noon, the clouds began to clear. We called ahead to Grand Portage along the Lake Superior shore and got a favorable weather report. Then Don and I flew in the Aeronca Champion, and Park Service ranger Dave Stimson flew in the Cessna with Art Tomes.

As we approached the island and the Champ lined up on its approach to Washington Harbor, I could see right away that there was far less ice around the island than last year. Ice covered the mouth of Washington Harbor. The west end of Siskiwit Bay on the south shore was frozen, but otherwise sheets of ice floated on the bay. Likewise, pockets around the rest of the island were frozen, but for the most part, the water was ice-free or contained floating ice we couldn't land on. That was bound to limit our options for checking out wolf kills from the ground.

The next morning, Don and I wasted no time in getting into the air. After breakfast, we made our emergency kits of sleeping bags, miscellaneous snacks, and ropes (for an unexpected tie down in case weather forced us to land away from camp). We tied snowshoes to the struts and topped off the gas tank.

We cranked the prop, climbed into the plane, skied down the ice of the harbor, and lifted into the air. We flew along the south

shore of the island to make a general reconnaissance of ice conditions. Less than an hour into the flight I was getting airsick again,
so we put down on Intermediate Lake, near the midpoint of the
island, so I could rest and let my head clear. An hour later, we
took off again and almost immediately discovered the large pack
of fifteen wolves on Lake Richie, presumably the same pack, with
many of the same wolves, we had seen last year. We continued our
survey, finding wolf trails—"abounding on the shores," I scrib-
bled in my notes. They really were almost everywhere—not only
the tracks of the big pack but also tracks made by what appeared
to be three or four wolves, and singles.

After lunch and refueling at Washington Harbor, we returned
to Lake Richie to watch the pack. They were still there, and this
time we spotted yet a sixteenth wolf, presumably Homer, the
lone wolf we had identified the year before. A pair of wolves were
hooked up in sex, and as we watched from above, it appeared there
might be two other breeding pairs in the pack.

We were curious whether the wolves, which had grown accustomed to the airplane last year, remained so. Don flew a couple of
low passes over the pack, and they showed no signs of fear or even
concern. So he circled again, and we landed about 150 yards from
the wolves and continued taxiing slowly down the ice. The wolves
were unmoved and unmoving. Only as we accelerated and lifted
off did they jump and scatter into the nearby woods.

An hour later we returned. Don set the plane down within 60
yards of the nearest animals. They stood for a few seconds and
then ran back and forth for a minute or more as I took movies.
Soon they scattered into the woods again. Minutes later we heard
a growing chorus of howls from the woods until it sounded as
if the entire pack was howling and yelping. The sound gradually
diminished, though there were still occasional single howls. After
waiting in the plane for a half hour, we took off again and discovered tracks that showed that the wolves had assembled on a small
knoll, where most of the howling probably originated. But the
wolves had vanished, and we weren't able to find them again.

* * *

To pick up the trail the wolves had followed during the night, we sometimes had to land the plane so I could follow the tracks on foot as Don circled overhead. On our second day of flying, we had lost the wolves' trail at Intermediate Lake. Don dropped me off, and I followed their tracks through the snow west up the Greenstone Ridge and then back south to Siskiwit Lake. Don picked me up, and we found the large pack crossing the snow-covered lake.

As we watched the wolves, we routinely flew ahead of them to spot the next moose they might encounter. Doing so allowed us to observe the behavior of both wolves and moose before and during the attacks. It also gave us the opportunity, if there were no moose in the immediate area, to dash back to refuel at the base at Washington Harbor or at our improvised fuel cache at Mott Island, or to quickly land on a handy lake and gas up with the five-gallon can we carried in the plane. This way we minimized the chance we would have to break off our observations in the middle of a hunt. We could see nothing was likely to happen as the wolves crossed Siskiwit, so we sped back to Washington Harbor to top off.

When we returned, we found the pack had crossed the lake into the woods on the southeast shore, climbed a ridge, and tracked northeastward. At 4:35 in the afternoon, we noticed the wolves had changed direction and begun running upwind on the open ridge toward a cow and two calves. The way the wolves had changed their course suggested they had caught wind of the moose from a full mile and a half away. The thought occurred to me that moose must give off a powerful amount of scent for the wolves to sniff them out at such a distance.

When the pack closed to within three-quarters of a mile, several wolves grouped up on a high ridge and pointed as if they were bird dogs. The moose now looked toward the wolves, apparently aware of their presence. A few wolves at the head of the pack charged down the ridge, toward the moose but a little to the north. Two wolves were far in the lead, and two other wolves ran south of the trail the moose had left.

The moose now were clearly aware the wolves were after them and ran toward Wood Lake. Encountering a steep cliff, they turned south. The lead two wolves gained quickly and soon caught up. As the moose ran through an open woods of young birch, a wolf ran along each side.

The cow ran close behind her two calves. Twice she feinted toward the wolves, which leaped out of the way. By now the rest of the pack was catching up. As the moose ran into a small cedar swamp, four or five wolves tore at the rump and sides of a calf and clung to it. Within 50 feet, the calf tumbled to the snow in a thick clump of cedars. The cow and other calf continued to run. Two wolves followed but quickly broke off the chase and joined the rest of the pack at the wounded calf. The cedars obscured our view, but the calf appeared dead within five minutes after it fell. The cow stopped running and started back toward the downed calf. But gradually, the cow and surviving calf drifted back toward where the chase had begun.

Don and I continued to circle the scene of the kill. Within a half hour, the wolves had torn the skin from the neck and left side of the calf's chest and had eaten the heart, parts of the lungs, the rump, and nose.

That was when I suddenly realized what a golden opportunity I had with this situation.

"Don," I blurted out. "I want to land and get a close-up look at the kill."

"You can't do that!" he replied. "It will be too dangerous with the wolves still feeding."

But I had my Park Service–issued revolver. And you already know the rest of that story.

Three days later we followed the pack of sixteen as it strung out single file and advanced slowly along an old wolf trail north of Moskey Basin. The wolves suddenly veered upwind and became alert, often walking briskly and then stopping and pointing as though scenting the wind. They bunched into a compact line and traveled three-quarters of a mile to within 250 yards of a cow and calf that were browsing directly upwind. The wolves turned into a thick spruce swamp, as though they didn't know quite where to find the moose. But then they turned again and made a beeline for their prey. As the pack closed to within about 100 yards, the moose began to flee, the calf ahead of the cow. The wolves continued to follow and soon were racing alongside and right behind the running moose.

As they ran, the cow stayed tight behind her calf, charging and kicking toward the wolves as they closed and tried to attack. From everything I had read, the moose is one of the most formidable and dangerous animals wolves can tackle. A couple of early observers I was familiar with had reported wolves that had been badly injured in pursuit of moose. So if they have the option of chasing other prey, they take it. But these wolves had no other real choice (except the occasional hare or beaver during the warm seasons). They were forced to prey on moose. Presumably, during the decade they had been on the island, the wolves learned to fear and avoid the adults' fearsome feints and kicks.

One wolf did manage to nip the calf on the rump, but the moose didn't falter. The chase continued for 200 to 300 yards, the mother sticking close behind her calf. But eventually, in the chaos of fending off the wolves and crashing through the forest and loping over rough terrain, a critical space opened between the two moose. Two wolves followed the cow, driving her farther from the youngster, but the rest of the pack mobbed the calf, snatching at its rump and flanks. One grabbed its left hind leg. The cow caught up with the group and stomped on one wolf, but the animal jumped up and appeared unhurt. Perhaps the snow buffered the wolf. The other wolves lost hold of the calf but continued pursuing it for another 100 yards before attacking again. They pulled the calf down and tore at it, but the calf struggled to its feet, and the cow rushed in again. Some wolves fled, but others attacked the cow, driving it away. Then the wolves lunged at the calf again. One grabbed its nose as three or four tore its neck and throat. Others bit its rump.

The calf's rear end crumpled, but it struggled forward, dragging its hind legs and several wolves with it. It stood once more, and the cow made another charge, but a single wolf chased it off. The wolves swarmed the calf, and this time it went down for good. The wolves crowded around it and began to rip it open and eat. The cow wandered off and eventually headed to where she and her calf had first been pursued.

Yet even as Don and I watched these examples of relentless and successful pursuits, we occasionally saw something different—and much more puzzling. Several days later, we thought we were watching another successful hunt unfold. We had tracked the large pack to Siskiwit Bay and up the Feldtmann Ridge Trail into Siskiwit Swamp, where they appeared to scent three adult moose 200 yards upwind and started toward them. As the wolves drew nearer, the moose took off, two to the west and the third to the north. The first two wolves in line pursued the two moose. They really flew and overtook the pair within a couple of hundred yards. They continued the chase for a half mile through thick second-growth birch but never appeared to attack.

We lost track of the two as we focused our attention on the rest of the pack, which had taken off in the other direction after the lone moose. They, too, quickly closed the distance and ran behind and alongside for perhaps 300 yards. Then the lead wolf slammed to a stop and actually lunged at the wolves coming up from behind to prevent them from continuing. The following wolves nearly tumbled over one another to get out of the way and ran up their back trail as the moose ran off through dense second-growth cover.

The wolves regrouped on a nearby trail. The two wolves that had chased the pair of moose returned. (We found the moose slowly walking away and never discovered whether the two wolves had simply given up the chase or whether the two moose had turned to face them and had scared them off.) At any rate, the pack rested for a few minutes in a nearby swamp. Then they retreated to Lake Halloran and rested another hour.

It wasn't that the wolves weren't hungry. From what we knew

of their habits and where they had been the last couple of days, we knew they hadn't eaten and were due to make a kill.

And just a little while later, they were at it again. They were walking along a ridge when they stopped about 100 yards from a cow and calf. The wolves were directly upwind of the moose, and though the pack seemed to catch their scent, they didn't seem to be able to locate the moose. The calf was standing, and the cow was lying down. But then the cow rose, and the wolves immediately shot after the moose.

The chase unfolded like many others. The calf ran ahead of the cow. The wolves closed the distance. The cow kicked and lunged at the wolves. But then the wolves dropped back. The moose slowed to a walk. And after a few seconds, the wolves stopped entirely. The moose stopped too and turned to watch the wolves. After a couple of minutes, the wolves simply left.

It was just a half hour later, as the sixteen wolves traveled west along the Lake Superior shoreline, that they suddenly veered inland, apparently following the scent of a lone moose on a ridge. The moose took off, apparently thought better of running, and lodged itself against a spruce and waited for the wolves to close the distance. As the wolves crowded toward the moose, it flailed with its hooves, seeming to land blows with its back feet. The moose successfully defended itself for several minutes but then took off along the ridge through a stand of spruce. The wolves pursued, biting its flanks and rump as the moose plunged over the edge of the ridge. As the moose slid into the creek bed, the wolves cascaded down the embankment and swarmed all over it. They bit its back and flanks. One grabbed its snout. The moose shook the nose wolf from side to side, as the other wolves continued to attack its hindquarters.

The moose continued to struggle for several minutes, but the wolves were packed around as if at a mess hall dining table. The wolf on the moose's nose held its grip. After about ten minutes the moose appeared to be dead, and the wolves began ripping it apart.

So what was the difference? Clearly the wolves were desperate to eat. In several instances they gave up the chase or never really took up the pursuit in earnest. Yet they pressed their attack successfully against the moose that had, to my eyes, put up perhaps the toughest defense. What had the wolves detected that I hadn't? What weakness did they exploit, and when did they realize they could make good on their attack?

We would encounter this puzzle again and again. One attack was successful. Then, in similar circumstances, the wolves seemed to give up without even trying. And we couldn't understand why. In one especially memorable case in late February, the pack was able to size up the situation within five minutes and realize that its quarry was invulnerable. That standoff lasted only long enough to give me the chance to capture it in a photo that would end up not only in *National Geographic* magazine but in many books and articles for decades to come.

I had learned to take 35-mm colored slides during my last year as an undergraduate and was currently wielding an Argus C44 camera as well as a Bolex 16-mm movie camera. Although I had snapped many routine photos during my first year on the island, I had longed for a chance to photograph the wolves hunting moose. But despite the many opportunities, circumstances kept thwarting me. Then suddenly, the large pack encountered a single, stubborn moose on an open hillside. I saw my chance.

"Don! OK if I open the window? I want to get some pix."

"Let me slow 'er down first," Don yelled, as he eased back on the throttle.

Don then displayed his awesome talent with the Champ. Slowly and smoothly the light aircraft circled over the incredible scene of an especially defiant moose standing off the entire pack. I slid back the Plexiglas window, snapped away, and managed to capture the few minutes before the wolves concluded they wanted no part of this moose, turned tail, and headed away.

The resulting photographs, which I finally saw when back on the mainland weeks later after they had been developed, were unique. The photograph captured fourteen of the wolves surrounding the moose. It really looked like the poor moose was done for. In reality, the huge creature's defiance had left the threatening wolf pack with empty stomachs.

That photograph turned out to nicely supplement my income for the next several years.

The winter was shaping up as a great winter to fly. The weather was mostly favorable. Early on, we surpassed the number of chases and successful hunts we had witnessed all the previous winter.

Yet there were still days we couldn't fly—it was too windy, or visibility was too limited by snow. Then we turned our attention to camp.

During the off-season, after the Park Service had closed up the premises and before we arrived, the deer mice had apparently interpreted human absence as an opportunity to plant their flag in the ranger station. They were everywhere—in the house, in the shop. On days flying was impossible, I indulged my interest in wildlife and trapping by laying out deer mouse traplines. I would rise each morning and check my lines, catching sometimes a single mouse, sometimes several. I would prepare study skins of my catches as a way to record for posterity examples of what was probably the most common mammal on the island. How or when the deer mouse (*Peromyscus maniculatus*) arrived at Isle Royale was anyone's guess. It was certainly small enough to stow away aboard a fishing boat or even an Indian canoe. Yet it might have washed ashore on some flotsam from the mainland. Conceivably the species might be one of the oldest residents of the island. I imagined that in addition to getting rid of household pests, I was also providing museum-grade specimens that might shed light on such mysterious questions.

Probably the second-most common mammal on the island was the snowshoe hare, and there was certainly evidence they were more plentiful than last year. Between the dock and our camp there were two or three well-used runs, where the previous year there had been none. Tracks peppered the lakeshore and formed trails all over the island. Wherever the snowshoe hare occurs, its populations run in boom-and-bust patterns—not just random fluctuations but cycles that last roughly a decade. In the days not too long ago when lynx were common on Isle Royale, snowshoe hares would have been their favorite food. In fact, because lynx eat almost nothing else, the population of hares and lynx would be closely synchronous. When hares hit a downward turn, the lynx soon followed. With lynx now gone, or nearly so, the island's red foxes were probably the hare's chief predator. Various raptors undoubtedly took some. Although some hare hair showed up in wolf scats, it was becoming clear from my scat analysis that wolves spent little effort chasing after animals as small as a hare.

Among the activities that had absolutely no relevance to our wolf study was the Park Service's ongoing battle against the aging and decaying structures on the island. Weather and our schedule permitting, Don Murray would fly chief ranger Ben Zerbey (who had replaced Dave Stimson after a couple of weeks) to some far-flung outpost of former habitation on the island, where Ben would take advantage of the low fire danger in winter to burn down old fishermen's shacks and unneeded Civilian Conservation Corps buildings constructed a quarter century earlier. In mid-March, he torched eight old buildings at the Siskiwit Bay CCC Camp. A few days later, Don flew him out to burn a couple more and then to Hay Bay to finish off some buildings at an old fishing homestead.

Certainly some of the structures had become dilapidated and perhaps even hazardous over the years, and the Park Service probably had little practical use for many of the buildings and little money or incentive to save them. Now that Isle Royale had been designated a national park, it seemed the Park Service was determined to return the island to a more natural state, even though humans formed a long thread in the fabric of the landscape. Immigrant settlers had fished the waters around the island for more than 100 years. Americans had been digging shafts and extracting copper (though in barely profitable quantities) for a similar amount of time. And, of course, Ojibwa Indians and their predecessors in the country had been hunting, fishing, pecking out copper nuggets, and living on the island for thousands of years before that. But with a nascent wilderness ethic in ascendance, dozens of fishermen's homes, mining camps, and CCC buildings were being erased to help restore the natural state. (It was a policy the Park Service would come to be criticized for in the decades that followed.)

But naturally most of our flight time was monopolized by wolf watching. And even when we weren't watching wolves hunt, we saw a few things that struck me as curious.

For one, among the large pack we no longer saw the three lanky, light-colored wolves that the previous year had rested more often and played more frequently than the others. We had guessed these were pups. Now, a year later, they had matured and were indistinguishable from the other adults. Simple enough. But the size of the pack had remained the same. Why hadn't a seemingly healthy pack of wolves that had exhibited plenty of mating activity and had demonstrated the ability to get ample food produced new pups? Was the presumptive breeding pair infertile? Had the pups died? Was some biological mechanism preventing population growth? We didn't know.

We did, however, still see Homer, the outcast wolf. He (or more objectively, it) continued to follow the pack, not mingling much with the others. We did see him hunt with the large pack, excitedly chasing after a moose that had gotten a long head start and was soon out of range. But on another occasion, when there was a lot of mating activity within the large pack—a time when hormones and competition seem to be at a peak—Homer would often run from the others and slink behind at a distance. It was difficult to describe the interaction between Homer and the rest of the wolves, except to say that there seemed to be stronger tolerance this year than last but still something short of acceptance.

We did get some inkling of how this rivalry could play out with neighboring wolf packs. In February, we watched all sixteen wolves, including Homer, chase a single outsider wolf at least a half mile across Moskey Basin. All ran extremely fast, but the lone wolf had incentive to run faster, and it reached the woods ahead of the others and continued at least a quarter mile without stopping. The big pack gave up when it reached shore, and the animals lay down and rested. The actions of the pack toward other wolves suggested that if they ever cornered an intruder, a mortal fight would ensue.

A couple of weeks later, the big pack traveled overland south of Ishpeming Point when half of the wolves struck out on what appeared to be a fresh wolf track. They followed it excitedly for about a quarter mile before returning to the others.

Another time, the big pack had chased a moose unsuccessfully and continued to Grace Harbor. While crossing the ice, the wolves continually looked toward Grace Island. From the plane, high above, we saw three other wolves bolt from Grace Island toward nearby Washington Island. Grace Island prevented each pack from seeing the other, but the sixteen animals kept looking toward the three, which were running and watching their back trail. The big pack continued across the ice. Several among the pack ran and chased each other, apparently play-fighting over a bone.

We saw this play-fighting at other times. The large pack had just left a kill and was traveling toward Cumberland Point when several wolves began chasing one another back and forth and around in circles, tussling over a bone. One animal would drop it. Another would pick it up and try to outrun the rest. The pack continued the game into the woods, as some wolves ambushed others. It was the kind of play anyone who had watched their golden retriever play with other dogs would recognize.

Since wolves had shown up on the island roughly a decade ago, we had all wondered how they got there. Wolves are good swimmers but not as aquatic as moose. A pair of wolves swimming 15 miles from the mainland seemed far-fetched. And except for the ill-fated translocation that brought Jim (aka, perhaps, Homer) to the island, there hadn't been any other efforts to reintroduce wolves. So most likely a pair or small pack had crossed the ice during a cold winter. And we had seen during our last two winters that wolves readily followed frozen shorelines and crossed ice-covered bays without hesitation. But watching the large pack venture out from the main island one day made me wonder how their immigration to the island had really unfolded.

One afternoon in early March, all sixteen wolves were loping northeastward down Rock Harbor. They had traveled 29 miles from their last kill. They reached Blake Point, the far northeastern tip of Isle Royale, and then headed straight across the ice toward Canada.

Soon they were more than a mile from land and began curving gradually eastward toward Passage Island, a rocky spit of land where a lighthouse stood, more than 3 miles from Isle Royale itself. Much of the ice was smooth, but in places, rough chunks of ice were frozen in a matrix of new ice. The pack appeared hesitant to cross from one texture of ice to another. The lead wolf seemed determined to forge ahead, but most of the pack lagged behind as if reluctant to follow. The lead wolf repeatedly backtracked to the wolves behind as if to urge them on.

I began to worry. What if the wolves left the island for good? What would that mean for my study? Yes, there were the few other small packs, but how much could I learn from those few?

"Don, I'm worried. What if they head to Canada? If they get much farther out, are you willing to drive them back for me?" I asked anxiously, still unsure about what I would want to do but at least checking my options.

This would be a momentous and ethically fraught decision that I would have to make spontaneously myself, without any way of reaching Dr. Allen or the Park Service.

Don didn't hesitate. "Certainly," he said.

Meanwhile, the wolves proceeded in fits and starts until they reached a stretch of shards of ice welded together. They stepped on it gingerly, and then they solved my dilemma. They turned around and headed back to the main island. I'm still not sure what I would have done, but my predisposition then was to have nudged the pack back to the island.

If the wolves had intended to reach Passage Island, I believed they would have set out on a more direct route. It really seemed as if they wanted to leave for the mainland. It was best for our project that they didn't have the nerve. They might have been able to cross all the way to Canada, but a few days later, the wind blew the ice and opened large fissures between shore and Isle Royale, which would have prevented them from returning.

March 4 turned out to be a field day, literally and figuratively, for watching wolves. Rising at 8:30, I checked my trapline (two mice!). The day was overcast, with a high ceiling, and after breakfast, Don and I lifted off to look for wolves.

Within a few minutes, we found them, traveling along the north shore of Siskiwit Bay. They turned inland toward two moose standing 150 yards upwind. The moose quickly realized what was up and ran. One skirted the shore of a frozen pothole, but the other avoided the lake and circled around until the wolves overtook it. The moose stopped running, faced down the wolves, and threatened them with its hooves. The wolves stopped, gathered together, wagged tails, and after a minute they left.

Still close to our base, we returned to refuel and picked up the wolves again northeast of Lake Halloran, where they appeared to be following a moose trail through a thick stand of spruce. As they approached two moose, the moose ran toward Siskiwit Bay. The wolves pursued, and the moose split up. The wolves all followed the smaller animal, which led them on a chase over small ridges and depressions and into the Siskiwit Bay CCC Camp, where the animal ran among the buildings. A few wolves ran alongside and behind the moose, nipping at its legs, but the rest of the pack lagged behind. As the moose cruised through thick cover and blowdown, the wolves fell farther off the pace. The moose ran through the Siskiwit Bay Campground and into Siskiwit Swamp. It stumbled over some downed trees and took cover in a stand of spruce. Four wolves remaining in the chase caught up and lay down near the moose, making no effort to attack. After a minute or so, the moose bolted. It ran for a mile through second-growth cover. But the wolves stayed where they were. After a moment, they gathered up and wandered off toward Lake Halloran. In all, the chase had covered well over two miles.

As they walked toward Lake Halloran, twelve wolves (the others hadn't yet caught up) stopped suddenly and then ran toward a moose standing 300 yards upwind. But the moose caught on quickly and trotted off, and the wolves quickly gave up.

The pack continued south of Lake Halloran and by late afternoon appeared to be tracking a moose. Actually, it was two moose, and the wolves walked right up on them but didn't attack. The two moose stood and looked at the wolves. The wolves looked at the moose. And then one of the moose took off. As the pack peeled off toward the first moose to run, the other bolted in a different direction. The wolves continued to chase the first moose and nearly ran into a third moose lying in an opening in the forest. The wolves immediately shifted their attention to this animal and surrounded it. The moose rose to its feet and confronted the wolves, which thought better of challenging it and returned to chasing their original target. But by now, the first moose was so far ahead the wolves quickly gave up. Two wolves, lagging far behind the rest of the pack, nearly ran into the third moose. But they decided not to engage and rejoined their fellow hunters.

The big pack reached the south shore and strung out on the Lake Superior ice. Heading southwest, they halted again and dashed inland toward a moose standing 75 yards into the woods. The moose ran slowly and then stopped near a small tree and turned to threaten the wolves. The wolves faced down the moose for less than a minute before turning and heading back to the shore.

I was so glad that I had finally overcome airsickness, because all the spinning around we needed to do to keep track of the ins and outs of each chase, each moose, and each wolf and to record all this on my data forms would have soon grounded me if I hadn't. Don's expertise with the Champ, his keen interest in all that was going on, and his stamina in continuing the chase also greatly contributed to the excellent data we were getting.

A quarter mile ahead of the wolves, standing 100 yards inland, were two more moose. They seemed to have detected the commotion from the previous confrontation and ran farther inland. When the wolves reached a point directly downwind, they turned and headed toward the moose but encountered yet three other large moose standing nearby. The three moose split up and fled, and the wolves scattered and chased all three. In less than a minute, the wolves had turned their focus to a single animal running parallel to the Lake Superior shore. One or two wolves managed to pull alongside, but as the chase stretched out to well over a mile, the moose climbed a small ridge, and the lead wolves let up. But the wolves in the rear took a shortcut and continued the pursuit. The leaders resumed the chase, but by now the moose was 100 yards ahead, and after a few seconds all the wolves gave up.

Toward sunset the wolves had nearly reached Rainbow Point. Again they veered inland, this time toward a cow and calf standing upwind. These moose stood their ground, the cow glued to the back end of the calf in her effort to guard it. The cow charged the wolves, and then both moose slowly ran off, the cow kicking and feinting toward the wolves as she protected the calf's rump. Even the calf charged the wolves. After a quarter mile, the wolves gave up.

The pack headed inland and by sunset had nearly reached the shore of Feldtmann Lake. Once again they appeared to scent something and headed upwind toward three full-grown moose. The moose all ran, but then one turned to face the wolves. The wolves passed on the invitation for a fight and instead chased the running moose, soon concentrating their efforts on one animal. Most of the pack fell behind, but several wolves kept up. But then the moose stopped to face its tormentors, and the lead wolves scrambled away. The moose ran again, but this time all the wolves simply watched as it disappeared into the forest.

By now the sun was well below the horizon, and the gathering darkness forced us home. All told, between watching wolves and taking advantage of lulls in the action to zoom home and refuel, we had been in the air more than seven hours. When Don and I finally landed, our seats sore, and tied down the plane, it was probably 7:30, and ranger Dave Stimson was standing by the dock, furious that we had stayed out so late without letting him know.

Durward Allen, in his book *Wolves of Minong*, described the scene better than I can:

The air was red, then blue, then purple, as Stimson gave them the most elaborate and competent "chewing" Don had experienced since Korea. I do not now recall what Stimson's role had been in the military, but Don judged that only long practice could produce the sophistication Dave displayed on that occasion. He was completely right, and the troops slunk up the hill with tails tucked and no arguments.

My efforts to record the wolves' travel over time was beginning to pay interesting dividends. By flying as often as we were able to and assiduously trailing and backtracking the pack of sixteen wolves to account for their whereabouts when we couldn't see them, we documented their entire travels for thirty-one days after the beginning of our winter season.

In all, the pack had traveled 277 miles, an average of 9 miles a day. But daily travel varied a lot. On most days, the wolves fed on kills and ended up traveling very little. On the nine days that the wolves didn't feed and seemed intent on hunting or covering ground, they averaged 31 miles. The longest distance they traveled in a twenty-four-hour period was about 45 miles—equal to the entire length of the island.

As we saw on the day we got our "chewing out" for showing up so late, the pack could put on plenty of mileage while hunting, especially if they had nothing to show for their efforts and had to keep looking. Unfortunately (from the wolves' perspective) they failed quite often. Sometimes it was a matter of bad luck—a moose got a head start, and the wolves had little chance of overtaking it. But we were beginning to see that many moose put up a formidable defense. Even cows protecting vulnerable calves were able to outfight and outlast a pack of sixteen hungry wolves.

In early March we were circling above the pack of sixteen as they traveled the south shore. Near Hat Island they suddenly cut inland. When they were 50 yards directly downwind of a cow and calf, they apparently were better able to locate the source of the scent and veered directly toward them. The cow leaped to the calf's rear, and when the wolves were about 25 yards away, both moose took off toward Siskiwit Lake. They didn't run fast. No doubt the cow knew that sticking close to her calf was more important than speed. Within moments, the wolves swarmed at their sides and heels.

As we had come to expect, whenever the wolves came within striking distance, the cow lashed out with her feet. With each attack, the wolves fell back but then closed the gap immediately. As the cow threatened the wolves within range, other wolves dodged in to try to bite the calf, but the cow charged them too, causing them to scatter.

As the cow fought off the wolves, she fell behind her calf by 10 yards. The wolves dashed in, but the cow refocused her attention and charged the wolves attacking her calf and drove them off. The calf, which seemed smaller than the other calves we were seeing that time of year, chased the wolves that got ahead of it.

This went on for about 2 miles. The moose stopped, and so did the wolves. The moose rested a minute and then took off again, but the wolves remained behind. After running about 150 yards, the moose slowed down, the cow now leading the calf. A few minutes later the wolves again gave chase. But their effort was halfhearted. They flopped down to rest as the moose ran on for a half mile.

A couple of days later, we saw more examples of the wolves' failure. And as had happened a week earlier when we spent so long in the air, the failure came in waves.

We spotted the pack of sixteen near Feldtmann Lake in early afternoon. They were crossing the wind when they spread out and ran excitedly back and forth. We suspected they caught a whiff of a cow and small calf standing 300 yards away. Our suspicions were confirmed when the pack turned upwind and began heading toward the moose. The chase was on, the cow tucked in behind the calf, and the wolves following alongside or strung out behind the action. On they went, through spruce swamps and alder swamps and stands of old white birch and aspen. As was so typical, we never saw the moose break into a wild gallop. They simply maintained their effortless loping gait, like trotters. Moose had been reported running anywhere from 19 to 35 miles an hour. Except through deep snow or blowdown, they couldn't outrun the wolves. But the thing is, the moose never stopped. After 3 miles without even having made a serious attack, the wolves simply fell behind and gave up, each wolf dropping to the snow and resting as the moose continued running for another quarter mile. The snow wasn't deep, probably less than a foot in some of the sheltered swamps the moose had traversed, but we suspected a light crust may have hampered the wolves. Certainly the moose's long legs and massive musculature gave them quite an advantage as well. Ten minutes later, the wolves picked themselves up, reassembled amid sniffing and tail wagging, and lay down for another ten minutes.

Less than an hour later, the pack was hunting again. Following a fresh track, they jumped three adult moose in a spruce swamp. The moose skedaddled, and the wolves chased one until it turned and confronted them. Then they chased off after one of the others. But this moose had a long head start, and the wolves stopped and wagged tails.

An hour later the pack was following a ridge near the mouth of Grace Creek toward two moose lying in the snow. Fortunately for the moose, the wind was in their favor, not the wolves'. The moose jumped up and ran, and the wolves never noticed until they stumbled across the track and slowly followed it. By now the moose had a long head start, and the wolves never took up the chase. I'm not sure the wolves ever saw them.

Soon after, the pack jumped four adult moose, which ran off in different directions. The wolves zeroed in on one, but the moose glided over extensive blowdown of mature trees, which seemed to tire the wolves. The moose stopped and charged the wolves, which scattered, and then resumed running. The wolves never really recovered and soon quit.

The wolves drifted back to where the chase had begun and scented and began chasing yet another moose. But after only 25 yards or so, the moose stopped and stood its ground. The wolves got the message and after a minute of indecision, walked away.

Given so many moose and the wolves' ability to travel so long and hard, it was beginning to dawn on me how the populations of these two adversaries were able to survive: The wolves ended up checking numerous moose, and eventually they would find one they could kill and eat.

Wolves weren't the only hunters on the island—or even the sole canid hunters. Coyotes had been here in numbers before wolves arrived, and we had reason to believe that competition and antipathy between the two led to the coyote's demise. Likewise, red fox had homesteaded here long before wolves showed up. The earliest report I heard of red fox on the island came from fisherman Pete Edisen, who remembered seeing wild red foxes hanging around the cages where Bill Lively raised black foxes in about 1925. (Black foxes, also called silver foxes, are a melanistic variant of red foxes that occurs in nature and are highly valued for their fur.) Biologist Adolph Murie, working on Isle Royale in 1929–30, also reported seeing a few red foxes.

Because wild canids of all species tend not to play well together, we wondered what kind of relationship wolves had with red foxes. Late in March, we got a clue.

We had been following the large pack near Feldtmann Ridge when the wolves began searching intently through a swamp southwest of Lake Halloran. The lead wolf took off, and about 125 yards ahead of it a red fox bolted away. The wolf ran to where the fox had been hiding, and a second fox flushed. This one was not so fortunate as the first, and within about 15 yards the wolf caught it in its jaws, shook it violently, and carried it beneath some trees, out of our view.

The wolves soon found a moose carcass that apparently the foxes had been scavenging. Whether the wolves had been following the scent of the foxes or the moose through the swamp, I couldn't tell, but I wanted to get a look at the dead moose. So we landed, and I hiked in to the carcass as Don lifted off and circled overhead. Don later told me that most of the wolves ran off when I approached to within about 150 yards. As I got to within 20 yards, two more wolves looked up from the moose carcass and slowly moved away. I took several more steps toward the moose, when the remaining four wolves spotted me, jumped up, and ran off. The pack reassembled about 150 yards away and lay down. Judging by the poor condition of the bone marrow in its femur, I guessed the moose had starved before the foxes found it. As for the second fox, it was dead with its guts spilled out onto the snow. None, that I could tell, had been eaten.

I rejoined Don. We lifted off and continued to circle the area. A half hour after I had left the carcass, two wolves tentatively walked back toward it. One wolf sniffed its own trail cautiously, suddenly seemed to get a whiff of something, ran back away from the carcass a few feet, and stood looking at the moose. Ten minutes later, when we finally headed home, the wolves still had not returned all the way to the carcass. Did I smell that bad?

The next day as we flew over, the wolves were feasting on the dead moose. We couldn't spot the dead fox. We didn't know if it had been eaten or simply carried away to where we couldn't see it. But another fox lay curled up about 100 feet away, apparently napping while waiting for its chance at a meal.

On March 17 we had one more opportunity to see what could happen when everything went right for the wolves.

We had been following the big pack near the Island Mine Trail when they picked up a scent, turned upwind, and ran into a thick stand of spruce. Two moose sliced through the stand and appeared to split up. The evergreens were dense, and it was tough to keep track of the animals running beneath us. Then a moose, a calf closely pursued by two wolves, ran from the cover of trees. We could only guess that its mother had become separated in their flight from the wolves and either couldn't find the calf or was too desperate in its own fight for life to rejoin it.

The calf ran down the trail and off into the swamp again. The wolves were soon nipping at its hind legs. Then one latched on to its rump and the other its throat. The calf trampled the wolf at its neck, but the wolf hung on as the calf stomped it and dragged it around. Finally it let go, but the other wolf still clung to its rump. The wolf that had released its hold stood on its hind legs with its front paws on the moose and chomped at its throat again. The calf brushed it off against a tree, but the wolf dove under the moose and grabbed its throat again, running along amid the moose's flailing legs. Two other wolves caught up. One bit the calf's head and then clamped onto its nose. The other bit the rump and hung on until the calf collapsed in a clump of trees. The wolves swarmed it, and within three minutes the calf ceased struggling. The rest of the wolves gave up chasing the cow and found their way to the dead calf.

Don and I landed south of Senter Point, and I hiked in to the kill. The wolves had run off. Most of the flesh was cleaned off of the head and throat and from the upper hind leg and pelvis. Part of a shoulder had been eaten. The abdomen had been ripped open, the intestines pulled out and partly eaten, along with the liver. Whether these parts are preferred or whether they merely represent the points of attack, I didn't know.

I hiked back to the plane, and Don poured coffee from his steaming thermos as we unwrapped our meat sandwiches and munched on cookies.

Don and I watched many more incidents of wolves pursuing moose than I have mentioned here. It truly was a banner year for wolf watching. I greatly expanded the observations and data I needed for my thesis and my own intuitive understanding of wolf and moose behavior.

Partly, Don and I were simply getting better at doing our work. By now we were expert at finding and knowing wolf and moose tracks. We had developed a pretty good understanding of the distances wolves traveled under various circumstances and the routes they commonly followed as they traversed the island. We had devised a routine for scanning ahead of the wolves to better anticipate when they might encounter moose and to plan our refueling trips accordingly. Don had developed an uncanny ability to circle an area and travel with the wolves as necessary to keep my eyes on the action at all times. He couldn't have done much better with a helicopter.

We were also helped out by the weather. Snow depth this year was only 12 to 16 inches on the flats, significantly less than the year before. How that impacted the wolves' success in dragging down moose wasn't obvious, but almost certainly they could travel more easily and faster, and that led to more encounters with moose.

But the biggest factor in our success was good flying weather. Owing to clearer skies and less wind, we spent half again as much time in the air as the year before.

Put it all together, and we logged thirty-five hours actually watching wolves hunt, compared with only nine the previous year. We saw thirty-three hunts involving sixty-six moose, compared with only nine hunts and fifteen moose the year before.

We were beginning to draw a bead on one of the more vital questions of this study, at least as it concerned these particular moose and wolves on Isle Royale: how often did the big pack kill a moose? So far, it looked like the answer was about one moose every three days.

We could also draw some generalizations about the moose that were most vulnerable. Clearly calves were most defenseless, even when closely guarded by a defiant cow, and the wolves seemed to work especially hard to separate a calf from its mother.

We also realized that a healthy, full-grown moose that chose, through calculation or instinct, to stand its ground and confront the wolves—even a pack of sixteen—stood an extremely good chance of surviving the encounter.

If that was true, why would any adult moose choose to run? Perhaps the runners were the weak and the old that feared they would fare poorly in a prolonged fight. Perhaps some individuals were simply predisposed to run. Others perhaps overestimated—if a moose is capable of such a thing—the head start they had. It was an interesting question we hoped to tease out the answer to.

Finally, we were getting a more nuanced understanding of the distribution and organization of wolves on the island. This winter we had seen what appeared to be four groupings of wolves: the pack of fifteen plus Homer, and much smaller groups of three wolves, two, and one. We spotted the pack of two three different times, and the pack of three five times—all on the same half of the island but never on the same day. So I suspected that the two wolves were part of the pack of three, and the single wolf was the third animal. Our estimate of the island's wolf population hovered at nineteen or twenty.

Our observations that winter served as further evidence that wolves were far-from-perfect predators. Time after time they tested moose and failed to bring them down. In many cases, they confronted their prey and simply gave up, somehow sensing the danger or futility of a chase. Their limited success made me wonder if wolves, as had been largely accepted, were able to control and stabilize a herd of animals as large and powerful as moose.

The fair weather also aided our efforts in counting moose the last two weeks of our stay on the island. Just within the area we had managed to survey the year before, we spotted two and a half times as many moose. While it's possible the moose population had jumped a bit, I'm sure the weather played a far bigger role.

But the Isle Royale winter wasn't going to let us off scot-free. By the end of the third week in March, we were packing our gear to return to the mainland. On March 21, the day we planned to leave, snowfall turned to a blizzard with gusts over 40 miles an hour. We hunkered down for the day. I took stills and movies of the thrashing trees and sheets of snow sweeping across the lake. When the weather cleared the next day, we flew back to Eveleth, and then I headed home to the cornfields.

ISLAND OF CHANGE

SUMMER 1960

≈

Although the second winter field season had been intense, fascinating, and very productive for my graduate research, I had been eager to get back to my family. While I was on Isle Royale, Betty Ann had taken little Sharon to her folks on Long Island, and it was great to finally see them again. We found temporary quarters for several weeks at Purdue and arranged our family housing for when we would return in fall. Betty Ann was pregnant again, and if all went well, was due to deliver at the end of summer, shortly after we planned to leave the island and return to Purdue.

This shuttling back and forth between new housing at Purdue and Isle Royale would take place thirteen times before I completed the project.

Meanwhile I tended to my notes, data, and administrative tasks and awaited the spring opening of the park so we could get there again as soon as possible. I was riding high because of how much good data I had been able to get. In fact, Dr. Allen had now decided that, instead of my working toward a master's degree, I should plan on going straight for a doctorate. I was elated.

So in early May 1960, Betty Ann, our eight-month-old daughter, and I drove from Purdue University north to Houghton, to start the summer field season. But snow and high winds blowing off Lake Superior were so fierce the *Ranger III* stayed in port. We spent the night at the Modern Travel Rest Motel in Houghton, a collection of small white lodges that resembled an army camp. The next morning we attended mass at St. Ignatius Loyola Catholic Church before boarding the *Ranger III* at 8:30 a.m.

One might say that attending mass had been prudent, because we soon witnessed the "wrath of God."

A 30-mile-an-hour wind was still howling out of the northeast and struck the *Ranger III* full force as the boat emerged from the Keweenaw Waterway to begin its 60-mile voyage across the open waters of Lake Superior. During the past two days, massive swells had built across the lake. The *Ranger III*, the new National Park Service boat, measured 165 feet long, 34 feet across. You might not think that waves on a mere lake could cause a 648-ton vessel to pitch and roll, but long before we reached Isle Royale, most all of the passengers were seasick, vomiting into buckets or over the rail, or sitting in silent misery. To try to minimize our seasickness, Betty Ann and I lay on our backs in our cabin, but periodically we had to rise to heat the baby's bottle and try to feed her. We took turns between vomiting spells to minimize the agony, but the only one of us who doesn't vividly remember that miserable trip is our daughter. She seemed to have a good time all the way to Mott Island, where we spent the night.

The next day the weather was even worse. Winds reached 50 miles an hour, driving rain and hail from the Canadian shore. Fortunately, narrow Rock Harbor was pretty well sheltered from the north. I helped a bunch of fellows from the Park Service launch their boat *Conrad L.* and then my own runabout, the *Wolf*. Once I got the boat running, Betty Ann, Sharon, and I motored over to our new quarters. Rather than house us in the old Holte cabin, halfway down the south shore of the island, the Park Service appropriated the Jack Bangsund house on Rock Harbor, across the channel from the Daisy Farm Campground and only a few miles from Park Service headquarters. Bangsund, a longtime commercial fisherman who had squatted on the site without ever owning it, had died the previous year.

We found the house, overlooking a fish house and dock on Rock Harbor. It was built of logs and later covered with tar paper. We looked it over, unloaded our gear, and got it ready to spend the night. The next morning, I woke up with laryngitis and a cold.

The rain had cleared, but the wind continued to blow at 50 miles an hour. We spent the next couple of days cleaning and fixing up the cabin and sorting dried foods that had been stored at Park Service headquarters, throwing out those raided by mice. A couple of days later, Bill Bangsund, nephew of the old fisherman, arrived on the *Voyageur* from Grand Portage to gather up some of his uncle's belongings and take them back to Minnesota.

We were waiting for the arrival of a gas stove and propane refrigerator, which in addition to my outboard motorboat, would be our most significant claims to modernity. But in the meantime, I needed to scavenge ice to fill our icebox. Each fisherman's home had the all-important fish house, a shanty attached to the dock that housed a fish-cleaning bench and a large bin of sawdust. In spring the fishermen would chop large chunks of shore ice and bury it in the sawdust. Then every few days they could cut pieces for their iceboxes. I was getting a late start on this—fishermen traditionally arrived when ice clung to the shores of the island and was easy to find close at hand. But it was mid-May, and much of the ice had already melted. I headed into Conglomerate Bay, a long protected inlet south of the entrance to Rock Harbor. Motoring to the head of the bay, I found ice in the shadows of the shoreline. I hewed off big chunks with a hand ax, loaded about 30 pounds into the boat with an ice tongs, and brought it home. I stashed most of it in the sawdust in our fish house, jammed a chunk in our home icebox, and took the remainder to our next-door neighbor, Pete Edisen, who had fished the island most of his life. The whole exercise made me feel like a pioneer and added to the charm of living on this wilderness island.

Within a few weeks, the Park Service staff and some of our helpful neighbors, including Pete, helped move the new stove and refrigerator into the house. We picked up groceries and mail regularly at the Mott Island landing and most Sunday mornings attended mass at Rock Harbor Lodge. We hosted my parents and sister from central New York State when they visited again later in the summer, picking blueberries, watching moose, and occasionally listening to wolves howl on the hill behind Daisy Farm. Such was the domestic life my wife and baby and I lived on our sheltered channel.

One of my most rewarding tasks was writing captions for an article that would be published in *National Geographic* magazine. Written by Dr. Allen and me, it described the wolves of the is-land and our work so far. I had taken courses in magazine writing and had been earning some nice pocket money, with twenty or so stories in magazines such as *Outdoor Life* and *Sports Afield*. But *National Geographic* was something special, notwithstanding that Dr. Allen was a well-known writer whose byline no doubt helped ensure publication, and the strength of our research pretty much sold itself.

All was good.

Except for the *Wolf*.

I was ever thankful that Bob Linn, the park naturalist, had lent me his runabout, which made it possible to traverse the island far more efficiently than I could have done exclusively on foot. But the reliability of the *Wolf* wasn't improving with its age. On occa-sion I would set out for some destination only to have the motor conk out, and I would drift listlessly on the waves until I could be towed ashore, although I was never stranded far down some lonely shoreline or left in a position where the wind could sweep me out to sea. I was always buying spark plugs, scrounging plugs, cleaning plugs, gapping plugs, changing plugs. And sometimes my amateur tune-ups worked—for a few days or sometimes only hours. One day the gas siphon sucked up water residing in the bottom of the gas storage tank on shore. That left me stranded until Pete Edisen towed me in. The housing on the motor cracked, and that entailed one of several trips to Frank Taddeucci, the park mechanic. I or-dered a new spare motor, just to be sure. And there were the nor-mal chores of emptying the boat, cleaning it, and hauling it up on shore to be washed, scraped, sanded, and painted.

The protected channel of Rock Harbor had long been home to about a dozen commercial fishing operations. The names changed, often through marriage, but the locations would remain. Among the longtime residents were Pete and Laura Edisen, who lived along the shore about a half mile northeast of us.

Pete and Laura would often stop by. A few days after we moved in, they picked up some of Jack Bangsund's belongings, which they had arranged to buy, and I helped them load the haul into their boat. Pete lent me a gasoline washing machine and helped me carry it to the house and get it running. We set up an account with Pete to supply us fresh herring for breakfast. When I stopped by, I would find Pete, easily old enough to be my father, in suspenders with a well-stained ball cap on his head, sometimes sitting on a stool, mending his nets, and spitting Copenhagen "snoose." Pete and Laura were salt-of-the-earth Norwegians, some of the nicest and friendliest people we had ever known. Before long, the Edisens were our most frequent dinner guests.

In late May, I hiked across the island to Chickenbone Lake to stash a sleeping bag in a waterproof bag and set up a food cache to sustain me during future excursions to the interior of the island. On the way out, I found morel mushrooms in abundance. Morels are easily identifiable spring mushrooms, one of the tastiest and most highly prized species in the north woods. I excitedly gathered them up and carried them back home.

The next evening we had Pete and Laura over for dinner, and I was eager to treat them to a dish of wild morels. Betty Ann fixed a wonderful meal featuring morel gravy and mashed potatoes. About halfway through dinner as we all chatted and savored the great gravy, I was suddenly horrified to notice in the gravy bowl several tiny white worms with little black heads. (I later learned that they were maggots from little flies that lay their eggs inside the morels. Experienced morel pickers know that one should always split the hollow morels and wash out whatever creatures are there.) There was really nothing I could do except carry on and not say anything, for we had all been thoroughly enjoying the gravy despite its, uh, additives. No one else seemed to notice, so I just said, "Pass the potatoes and gravy," and had some more.

I'm not sure if Pete ever caught on. If he did, he soon got me back.

One day Pete came over with something for me to try. He had been lifting his fishing nets and found a loon entangled in the mesh. The poor bird had drowned, and here it was, in its razor-sharp black-and-white beauty, dead but still fresh. Pete had no use for it and suggested I cook it up. I might have told him that I had practically lived on roadkill as a penny-pinching undergrad at Cornell. I don't recall if he ever said that he himself had eaten loon. I suppose the fact that he was giving it away did not strongly recommend its palatability. At the very least, here was an opportunity to try something new.

So Betty Ann boiled it up for dinner. It was what you would expect of a large, oily bird that had spent its life eating fish—but worse. Anyone who has tried to cook mergansers or northern shovelers might have a good idea. It was honestly one of the very few wild foods I could not choke down. Next time I saw Pete, I told him, "Please, don't bring me any more loons."

As I said, Rock Harbor, stretching all the way out to the northeastern end of Isle Royale, had been the center of a lively commercial fishing community during the late 1800s and the first half of the 1900s. There were several such enclaves around the island. About a dozen families, including the Sivertsons and Eckels, lived and worked out of Washington Harbor at the southwest end of the island. The Rudes, John Skadberg, the Holtes, and a dozen others encamped around Siskiwit Bay. Along the island's hard north shore, where cliffs drop steeply into the icy lake, a few fishing families found shelter in Todd Harbor, McCargoe Cove, and Amygdaloid Channel. All told, during fishing's heyday before World War II, about fifty fishing families operated in Isle Royale waters, many of the operations going back several generations. Some owned their land; most were squatters. They came out to the island in mid-April, when the lake was as clear and cold as vodka on the rocks. They retreated back to the mainland by mid-November. Some tried overwintering, but in the cold and snow, without their community around them, few made a habit of it. When the sea lamprey appeared in Lake Superior, the bottom fell out of the lake-trout fishery. By the time I showed up, the writing was on the wall. The Park Service was taking over properties where fishermen had squatted without title to the land. A few old guys stuck to their meager and dwindling livelihoods, continuing to fish for whitefish and lake herring (also called tullibees or ciscoes). Of course, no youngsters were getting into the business.

Yet I never heard the fishermen complain. I got to know quite a few of them, Pete Edisen best of all. I enjoyed talking to them. Most loved telling stories of their decades on the island. They were a link to an Isle Royale that was quickly vanishing. More to the point of my research, they gave me insights into a kaleidoscope of changing animal communities. None, to my knowledge, had a degree in biology, but they were outdoorsmen and keenly aware of the natural world. Without exception, they had spent more time on the waters and in the woods of Isle Royale than anyone else on the island.

So when I traveled around the island in the *Wolf*, I often dropped in on them, sometimes just for conversation and a quick bite, sometimes with notebook in hand for a more formal interview.

The Edisens, who had been on the island since 1916, told me they had seen a band of fourteen to sixteen caribou on the ice near Daisy Farm back in 1922. Pete had also spotted one of these "gray ghosts," as they have been called, in Conglomerate Bay. They said lynx were reported to be common before they arrived, but they themselves had seen none. Nor had they heard of any being trapped after they arrived. Pete told me hares were plentiful when he first settled, and remained so until the early 1930s. By 1939, he saw very few. He first began seeing moose sign in about 1925.

One day in late June I took the *Wolf* with Phil Shelton, Durward Allen's new graduate student, who was studying beavers on Isle Royale and had arrived a couple of weeks earlier, to the other side of the island. We picked up Bob Janke, the park naturalist. We spent much of the morning poking around the bays, points, and islets on the northeastern end of the island. It was a clear summer day, flat calm in the morning, the most carefree and intoxicating weather Isle Royale had to offer—the promise of warmth and clear sailing. We could see far down through the water to the green boulders and rocky ridge reaching far into the lake, out of sight. We continued around Blake Point to Crystal Cove on Amygdaloid Island, where Milford and Myrtle Johnson lived and fished, after moving from Star Island. (The Park Service destroyed the Star Island homestead, where Milford had fished with his brother Arnold since the mid-1920s, after Milford and Arnold gave up their fishing rights in the area.)

The Johnsons told me they saw caribou on Isle Royale in 1925. They also spotted lynx and coyotes. They told me Bill Lively was trapping coyotes for the bounty back then. They said they had seen many mink, especially around their fish house, but that in recent years they had become scarce. Last fall, they related, something had bitten into and chewed several whitefish, each 3 to 4 pounds, caught in a net set near the mouth of McCargoe Cove—just as an otter would do.

Milford also told me something surprising. He spent three winters on the island in the 1920s and in 1931–32 and saw tracks of a single wolf each winter. They had heard from a man named Cross, who ran a fur farm on Silver Islet, at the tip of the steep cliff-bound peninsula that juts toward Isle Royale from Canada. Cross said he watched wolves cross the ice to Isle Royale and return. It happened so often, he said, he could almost predict when the wolves would travel. (Cross's son later wrote me that most of these wolves were of the "smaller or coyote variety.")

Later in the summer, Phil and I motored down the south shore and dropped in for lunch with Sam and Elaine Rude at Fisherman's Home Cove. Sam said when he began coming out to Isle Royale in 1911, he saw a lot of hares on the island. He noticed signs of lynx as well but no coyotes. By the 1920s, the lynx were gone, but coyotes had taken up residence. Hares went through periods of abundance and scarcity. Sam first saw beavers in 1927. They became more numerous during the next several years. He told me that in the summer of 1948 he found a paw print much too large for a coyote on a beaver dam on Hay Bay Creek. Sam said he had seen enough coyote and wolf tracks to know the difference. Once wolves appeared, beavers became scarcer, and coyotes disappeared.

Unfortunately, this font of Isle Royale lore from the early 1900s was quickly disappearing. As commercial fishermen retired, moved, or died, the Park Service was busy eradicating the fishing culture, creating the look of uninhabited wilderness—an illusion I was actually quite fond of. But in fact Isle Royale had been fished, hunted, and camped on for thousands of years. During the summer, I spent several hours talking to Tyler Bastian, an archaeologist, and his four assistants, who were camped at Daisy Farm and even spent a day digging in the yard around the Bangsund cabin. Bastian was investigating the hundreds of copper mining pits hand-dug by Indian people during the past six thousand years.

Just as the commercial fishermen were passing from the scene, other inhabitants had come and gone—caribou, coyotes, lynx, and Native miners and hunters. What I saw at Isle Royale was a slice in time, not Isle Royale as it had been since time immemorial, but a snapshot of an island in flux from the time glacial ice retreated ten thousand years ago.

Meanwhile, I continued my usual summer fieldwork, hiking the island, bagging wolf scats, and collecting moose mandibles when I found them or someone brought them to me. The field season had barely begun when, camping at the Lake Desor lean-to, I stood up from a squat, felt a piercing pain shoot through my back, and found I couldn't straighten up. I crept over to a sapling and cut a walking stick, which I used to pull myself upright. I tried various stretching movements and exercises to loosen my back, all to no avail. So I sat next to my pack, pulled it onto my shoulders, raised myself with my new cane, grabbed my record player with wolf howling records, and hobbled 3 miles to Ishpeming Point, where I stashed the sound equipment in the Park Service radio box. My load lightened, I limped another 5 miles down to Crow Point, where ranger Bob Johnson, the new ranger at Malone Bay, met me as planned. After dropping him at Malone Bay, I boated the 17 miles back to our cabin and headed to bed. Day by day, my back slowly improved. But I would have relapses, and concern about injuring it again dogged me as I hiked that summer. Was that Cornell doctor right, that I should be planning to spend my career at a desk? No, because I loved it out here too much.

As I traveled the island on foot or by boat, I often saw moose swimming across bays and channels. Once, as I motored toward Daisy Farm to pick up Phil Shelton, I spotted a bull crossing Rock Harbor. So I sped home and picked up Betty Ann and a movie camera. We raced ahead to get Phil. I also radioed Ed Fuller, a moviemaker who was staying on the island. We all motored back out to the moose and took movies and photos until it clambered onto shore and vanished into the woods.

I suspected that a moose in deep water would be largely invulnerable to wolves, and I heard a story from a longtime visitor to the island that seemed to confirm that a moose would take to water to escape from wolves, something we wouldn't get to see in winter when the lakes were frozen. She told me she heard splashing in Mott Bay near Gale Island. She saw a moose about 20 feet into the water—"prancing and snorting," she said. A wolf stood on shore and after a few moments ran into the woods. The moose remained in the water for several minutes before it too strode into the forest.

Yet retreating to water wasn't a sure thing, as I discovered later in the summer as I hiked the Huginnin Cove Trail. I had been told three people had seen wolves there a few days earlier, so I was looking for sign. I stopped along an old beaver pond, where I found the remains of a moose calf, much of it already eaten. The tracks and matted grass around the carcass suggested wolves killed the moose in shallow water. Apparently, the calf sought protection in the pond but never reached deep water.

On several of my hikes this year I lugged the phonograph to play records of wolf howls to see what response they might elicit. And sometimes it worked! The yaps and howls from my record player vanished into the silence of the forest and then, seconds or even minutes later, a medley of yips, barks, and mournful howls

120

of ever-changing pitch floated out of the woods. Sometimes the response lasted only seconds—a couple of quick yips or a single howl—but sometimes the chorus lasted several minutes. True to the stereotype of a wolf howling at the moon (years later proven erroneous by one of my own graduate students), I succeeded at eliciting howls only in the evening, though I tried many times to evoke a response throughout the day.

Exactly why wolves choose to howl had been an enduring wild-life mystery. Adolph Murie described several times when wolves howled before leaving the den to hunt. He ascribed the behavior to restlessness. Writer and naturalist Lois Crisler believed howling was an emotional expression. "Like a community sing, a howl is not mere noise, it is a happy social occasion. Wolves love a howl," she wrote at about the time I began my research. "Some wolves love a sing more than others do and will run from any distance, panting and bright-eyed, to join in, uttering, as they near, fervent little wows, jaws wide, hardly able to wait to sing."

In my own work, I noticed that wolves did not seem to howl when they were actively pursuing prey, though we did notice as we circled above in the airplane that the large pack on occasion would howl *after* a chase, when the wolves were scattered over a wide area. Perhaps howling, among other things, served as a call to help them reassemble.

One evening before sunset, I played my records on a ridge near Hatchet Lake. Hearing nothing, I moved a mile west along the ridge and tried again. Silence. After several minutes I packed up my equipment and started down the trail toward the Hatchet Lake patrol cabin. Then I heard a low howl, repeated several times. I broke out the phonograph and played another record. The same low howl came back to me, and several high-pitched howls joined in. I guessed the wolves to be about three-quarters of a mile away. The ruckus continued for about ten minutes.

Another evening I was at home when I heard a pack begin to call across the channel near Daisy Farm. They howled for nearly two minutes straight. After they stopped, I set up the phonograph and played a record. The wolves issued a chorus. So I played another record. I heard just barks or short howls as the pack moved down the trail toward Moskey Basin. I ran down to the dock and fired up the *Wolf* to try to intercept them at an area where the trail crossed an opening where I might see them. But they never showed.

Far more often than not, my records were answered with profound, unrelenting silence. I never knew, of course, if simply no wolves had heard them, or if they chose not to answer. And if they chose to keep quiet, why—whether they were not taken in by the act or had some other motive.

Sometimes the response I got was better than I might have hoped for. One evening on Feldtmann Trail, just as the sun neared the horizon, I played the record and waited. Suddenly a bull moose burst out of the forest and stopped on the trail about 25 yards away. I played the record again, and he took off.

A few days later I played my records along the McCargoe Cove Trail. No wolves responded, but within a minute, five ravens flew above me to investigate. I played the records several times more before retreating to my tent at the McCargoe Cove site, where Phil Shelton, Durward Allen, Lee Smits, and I were camped. The next morning we were lugging our gear down to the boat when we saw that a yacht had also been moored at the dock. The folks at the yacht were all excited to tell me what had happened the night before. One camper said four wolves had attacked a bull moose, which he could "almost hear breathing" in front of his lean-to, and that one wolf was wounded and calling for help. We asked him how he knew this. Did he see them?

"No," he said. "We could tell by their howls!"

Lee Smits was the unwitting dupe of one of my favorite howling incidents. Lee, by this time in his nineties, was an ardent wolf lover and proponent of Isle Royale National Park. Yet for all the time he spent on the island, and all the love and reverence he had shown toward its largest predators, he had never managed to hear a wolf howl. One evening Lee stopped by our cabin for dinner. After, we put him up in an old "guest cabin" on the property. Late that evening, after Lee had turned in, I set up my phonograph on the dock to troll for wolves. Getting no response, as usual, I went to bed as well. In the morning, Lee greeted us with the exciting news that after all these years of visiting the island and searching for wolves, he had finally heard them howl! His dedication had been rewarded.

Of course, I didn't have the heart to tell him the truth.

One of my objectives in calling for wolves, besides a sort of gleeful desire to interact with the animals, was to try to better understand when and perhaps why wolves chose to interact with a neighboring pack. Another reason was to try to find the ever-elusive wolf den on Isle Royale.

Ranger Bob Johnson and I took the Park Service fire boat at Siskiwit Lake to check on the old den Roy Stamey and I had found a year ago on the north shore of the lake. The old den had caved in and had not been redug. The dirt showed fox tracks but none of wolves. I checked the den later in the summer. One entrance remained caved in, but it appeared another tunnel had been dug. I found a single, small track that may have been made by a pup or a fox. But, still, the amount of sign wasn't encouraging and suggested the den hadn't been used recently.

Finding an active wolf den would have been sensational for the research, so I did everything I could think of to locate one. Perhaps by playing the howling records in enough locations, I might get denning wolves to betray their location, so I tried that. Canadian wildlife biologist Douglas Pimlott had just published a paper in which he had used tape-recorded wolf howls and his own voice to elicit calls from the wolves and locate their dens. So during the summer, I broadcast my recordings at thirty-four locations around the island. In all, I heard just four replies—and three were from the same location on the same evening. Not as big a sample as I might have hoped. I searched both areas thoroughly but found no evidence of a den. Quite likely the wolves had merely been traveling through the area. My search for a den would continue.

One morning in late August, I was visiting at Pete and Laura Edisen's cabin, where the *Voyageur* from Grand Portage was

123

moored at the dock. Some of the campers on board told me they had seen a badly wounded moose near the end of the Chickenbone Lake portage trail. It had a large open wound on the left hind flank and didn't want to move.

I made arrangements to fly over the site that afternoon in the Park Service's Cessna 180. We circled the area but couldn't spot the moose. It was too windy to land on Chickenbone or nearby McCargoe Cove, so we flew back to Rock Harbor Lodge.

That evening I dropped in on Ben Zerbey, the chief ranger. Ben, as it happened, had just returned from Chickenbone Lake, where he had seen the wounded cow moose, still alive. He had found a couple of bloody beds nearby, but the moose had remained where he spotted it and appeared unwilling to rise to its feet.

The next morning Phil Shelton and I moored the boat at Daisy Farm, hiked the 8 miles across the island, and soon discovered a dead moose with a deep gash on its left side. We took movies and photos and performed a cursory necropsy of major organs. As we were working, two U.S. Fish and Wildlife Service biologists hiked by. They said they had seen the moose alive the previous evening, but that early this morning they had scared several wolves from the carcass.

In addition to the ragged eight-inch gash on the upper left hind leg, there were wounds on the throat and left cheek. Feeding wolves had eaten through to the body cavity, and a few loops of intestines hung out. Judged by wear to its teeth, the cow was probably six to ten years old. Her lungs were packed with fifty-seven hydatid tapeworm cysts, the most heavily infected moose I had seen. Undoubtedly such a parasite load had impacted her breathing and stamina.

The next morning Phil and I flew over the site in the Cessna and spotted at least three wolves near the carcass. We landed at McCargoe Cove and hiked in to a point across from the moose, where we scanned the scene with binoculars and saw that wolves had eaten little of the carcass since we first saw it. We heard wolves howling some distance to the southwest.

The next day I gathered up my overnight gear, bid goodbye to

Betty Ann, piloted the *Wolf* to McCargoe Cove, and hiked back in to my observation site. For several minutes I watched through the binoculars as a wolf fed on the carcass. It was quite a thrill to be observing a wolf at the carcass from the ground—much different from watching from the air as wolves fed. It soon disappeared, but then I heard howling off to the west, which continued for several minutes.

I hiked to the campground, set up my tent, ate a snack, and that evening hiked back to the kill. I climbed into a dead tree a few feet away and waited. At 9:30 p.m. at least four wolves began howling about 200 yards away. Howling continued off and on for the next half hour but, to my chagrin, gradually became more distant. The wolves knew I was there and were howling out of something resembling frustration. After waiting nearly two hours, I hiked back to my tent in the dark.

I checked the carcass the next morning and spotted a wolf crossing an open ridge about 100 yards from the moose. I made a last check of the carcass, took notes on the portions that had been eaten, snapped a few photographs, and headed home.

As I looked for wolves and wolf sign, I happened upon other Isle Royale wild residents as well. One of the most thrilling incidents happened early in spring, when the island's birds were nesting and their hormones had programmed them to be highly territorial. Walking from Lake Richie, where an old beaver pond was slowly draining and converting to meadow, I heard high-pitched screams overhead and looked up to see a peregrine falcon harassing a bald eagle. The eagle was young, still in its brown plumage. The falcon climbed above it, folded its wings, and stooped at high speed down on the much larger eagle before flying away. Both of these top-of-the-food-chain predators were increasingly rare around Lake Superior, as we would later learn, because of the accumulation of the pesticide DDT, which caused thinning of their eggshells.

I continued to find tantalizing signs of otters. The skipper of the U.S. Coast Guard cutter *Woodrush*, a 180-foot buoy tender on Lake Superior, told me that he and his fourteen-year-old son had watched a large mink-like animal swimming in Washington Harbor only about 75 feet away. He said they had been watching beavers earlier, and this certainly was too long and lithe to be a beaver, and yet it was too long and wide to be a mink. They got a good look, too, watching it for about ten minutes.

Later in the summer I found the tracks of a weasel-like animal (five short toes) at the outlet of Hatchet Lake. Measuring 2 inches long by 2⅜ inches wide, the tracks couldn't belong to anything but an otter. Though I never confirmed it by seeing one myself, I felt sure otters had taken up residence on the island.

Sharp-tailed grouse were having a banner year. As I walked the trails, I seemed to kick them up everywhere.

In the spring, during nesting season, I was heading toward Chickenbone Lake when a grouse near the trail put on her broken-wing act, running to lead me away from her nest. It didn't work; I found the nest under a ground-hugging spruce. I counted seven eggs. When I passed the nest on my way home, I checked again: now there were eight.

By late June I was flushing hens with chicks. They didn't appear able to fly, perhaps because they were so young. Three weeks later the chicks would flush and fly with the hens.

These brushland birds had been spotted on the island since 1905. On the mainland, they had become more common as loggers and fires cleared forests, which grew back as young aspen and open brushlands. That seemed true on Isle Royale as well, as sharptails seemed to become more common in the wake of the great forest fire of 1936. (Several decades later, long after I completed my project, sharptails vanished as the burned-over forest aged and created a fuller canopy. The last sharptail on Isle Royale was reported in 1986.)

Snowshoe hares were having a great year, too. Hares, like several boreal species, go through roughly ten-year cycles of abundance, for reasons we don't really understand. Certainly the hares of Isle Royale underwent periods of boom and bust. This year it was boom. I saw them everywhere, singly and in groups. Sheltering from the rain at Lake Desor one day in the spring, I watched four hares right in front of the lean-to. They were slap-boxing with one another. It was late May, near the peak of breeding season, and they may have been males, amped up on hormones, competing for females. One might even have been a female warding off suitors.

I have an especially fond memory of one of the hares. When the weather allowed, I enjoyed sleeping out under the stars. I could get away with this only in early summer before the blackflies and mosquitoes hatched. I always slept near one of the park's primitive lean-tos, so if it began to rain, I could take shelter. One especially nice night I settled into my sleeping bag near the Lake Desor log lean-to, with my backpack and clothes safely inside. Sometime in the night, I awoke on my back with a funny feeling of slight pressure on my chest. The stars were bright, and there may have been some moonlight as I tilted my head to look down at my chest. There, in total comfort, rested a full-grown snowshoe hare, itself settled in for the night on what must have seemed to be one of the softest, warmest, coziest logs it had ever lain on. My slight movement startled the creature, and off it hopped, never to enjoy such a comfortable log again.

By the end of August, we were already making plans to leave the island. Phil Shelton and I made one more trip—this time by air—to McCargoe Cove to visit the moose that had been killed near Chickenbone Lake, in hopes of getting one more glimpse of wolves. The plane dropped us off at the dock. We then hiked in to our observation site and watched for the last three hours of daylight, woke the next morning, and watched from our lookout for another three hours. No wolves. After collecting specimens from the well-eaten moose remains, we hiked the 10 miles across the island to Daisy Farm and caught a boat ride home.

The next day we packed, and the day after that Betty Ann, Sharon, and I boarded the *Ranger III* back to Houghton. Even though a rough northeast sea followed us back to Michigan, we were thankful our return was nothing like our trip over.

PUTTING TOGETHER THE PIECES

WINTER 1961

❧

Flying in to Washington Harbor for our 1961 winter field-work, we were met by a highly exercised moose.

I flew over to the island with Don Murray in the Aeronca Champion. Dr. Allen and ranger Roy Stamey were out ahead in the Cessna 180 and had already set down on the ice in front of the Windigo ranger station. As Don and I circled to land, I spotted a moose under a balsam, maybe 20 feet from the back door of the cook shack. Despite the commotion of the planes and the humanity spilling out of the Cessna, the moose didn't flee. Instead, it seemed to me to be wounded or in some kind of distress. Once we slid to a stop, I met up with Dr. Allen. This was his first winter with me, and I was anxious for him to witness some of the wondrous sights I had viewed each winter. We strapped on snowshoes and shuffled up the hill to see the moose.

We sneaked behind the cabin to stay out of view until the last moment. As we peered around the corner of the building, we saw a mature bull standing beside the heating oil tank-trailer. He had apparently jousted with a heavy-duty extension cord that had hung coiled on the side of the building and was now entangled in his antlers and wrapped around his head and shoulders. As we stepped into the open, I walked slowly toward the moose and took photos. At about 25 feet, the moose lowered his head, flattened his ears, raised his hackles, and charged at full gallop. Luckily for me, for he had caught me flat-footed, he came up short against the end of the extension cord and stopped as though he hit a wall.

We couldn't figure out how to free the moose—or use our

back door. So for the time being, we left him alone to help unload gear from the planes.

Don and Roy started the Studebaker Weasel and began transporting equipment and supplies to the cabin. It was still about zero, and they had to heat the generator with a blowtorch to start it. We hauled in food from our root cellar, waded through the snow to the cabin's front door, turned up the heaters, and settled in with our gear. We cooked and ate dinner and made plans for the next day. Meanwhile, the moose spent the night by our back door.

The next morning, Dr. Allen and I took turns flying with Don to check out a kill we had spotted near camp as we flew in. Doc and I then snowshoed to the kill. He was impressed with the blood and gore, which was old hat to me, but his main interest was the possibility of setting up a remotely controlled camera he was anxious to try out.

After lunch we began discussing what to do with the tangled moose. We all felt sorry for the animal, yet no one volunteered to untangle him. He was full-grown, healthy, and probably weighed close to a thousand pounds, with a full set of antlers—still solidly attached to his skull. Unfortunately for the moose, we decided he was destined to become Dead Moose 74 in our collection—and also an ample supply of steaks, burgers, moose loaf, and moose balls.

The only gun in camp was the .38 Special revolver that I was supposed to carry when I checked on wolf kills from the ground, severely underpowered for an animal the size of a moose. We tried our best, but we had to reload twice, ultimately firing seventeen shots before the beast succumbed. We necropsied the huge creature in the afternoon and then butchered him. Dr. Allen, the amateur chef that he was, was elated.

The next day, we hitched up the Weasel to the remains of the moose and towed them to a point on Washington Creek, near a lean-to and makeshift blind where we hoped we might photograph wolves with the remotely controlled camera.

My plans this winter were the same as before: fly every day weather permitted, follow the wolves, watch their pursuit of moose, and study their interactions with other wolves, both in their own pack and with other packs. As before, I would snowshoe in to kills to gather what information I could about the age and condition of the victims.

One difference this year compared with the previous two winters was the snow—there was more of it, a thick, wet blanket 2 feet deep. Slush was evident on most inland lakes and bays, where the weight of the snow displaced the water underneath so that it seeped up and flowed over the ice, never quite freezing under the insulating blanket of snow. The conditions affected not only our fieldwork but also the wolves and moose—primarily to the detriment of the wolves.

My first encounter with the slush did not go well. Don had dropped me off on Lake Harvey to plow about a mile in to a dead moose. But the deep, sticky snow made snowshoeing tough, and I stumbled and wrenched my back again. I managed to get back to the lake, where Don picked me up, but I was laid up in bed for the next five days while Don flew with Dr. Allen and Roy Stamey.

The slush also created a trap for the plane. On a couple of occasions, Don and I touched down on snow-covered ice, only to feel the plane shudder as the skis sank into the wet snow, so Don would open the throttle and lift off again. At least three times after Don landed the plane, we didn't have enough power to break free of the slush to take off. So we resorted to what was apparently an old bush-pilot trick of cutting brush to lay along a short takeoff ramp. We maneuvered the plane's skis onto the brush to gain enough speed to break free of the snow. Once even that didn't work. So Don and I slid the plane back onto the brush ramp and snowshoed a runway. Finally we got back into the air. Oh, my aching back.

I did eventually benefit financially from that unusual ploy: *Flying Magazine* readily bought my article, "Skis, Slush, and Snowshoes," complete with photographs.

The conditions hampered the wolves as well. Several times we watched wolves struggle in the deep, wet snow.

During the past two years I had grown accustomed to watching

131

the large pack of wolves follow shorelines to put on miles. But this year, the wolves seemed to spend more time in the woods, paralleling the shore at a distance, as if they found the going easier there. It was hard to tell why—whether they found crusty snow that supported their weight or fluffier snow that allowed them to plow through it more easily. Maybe they were simply put off by the wet slush they found on the lakes.

One day we followed two wolves and watched them take turns breaking trail through deep snow. When they reached a steep escarpment near the Greenstone Ridge, they simply took long leaps and plunged down to a lower elevation. A few days later, we spotted the large pack trudging out to Gull Rocks, a cluster of outcrops about a half mile off the north shore of the island. They rested often and had a hard time struggling through the wet snow.

Several times we watched as wolves picked up the chase of a moose. On its long legs, the moose would stride through the snow, unimpeded. But the wolves would flounder in deep drifts, fall behind, and soon abandon the pursuit. In one instance, the wolves had actually grabbed a moose by the hind leg, but the moose kicked free and was able to put enough distance between itself and the pack that the wolves tired and gave up. Deep snow seemed to favor the moose.

Despite the slush on the lakes, the ice was more extensive than in years past. All the harbors and bays around the island, including big windswept Siskiwit Bay were frozen over. The previous two years the ice between Isle Royale and the mainland was tenuous and shifting. Ice often appeared to connect Isle Royale with Canada, yet after each high wind, the ice span disappeared. But this year, despite several windstorms, the bridge to the mainland remained intact from mid-February through our entire field season.

Nevertheless, some wolves, some of the time, still distrusted the ice. One evening, Don and I watched seven wolves trekking northward from Hay Point across Hay Bay. Most of the ice was bare, and the wolves seemed reluctant to walk on it, perhaps because bare ice is often new and thin. They tried to walk on chunks of old, snow-covered ice as though leaping from rock to rock across a stream. And when they reached the end of the snow, they followed the milky-colored cracks, as though they found the opaque ice more trustworthy.

One wolf wouldn't be reassured. Though it was tempted and made several hesitant attempts, it refused to follow the pack across the snow-free ice. Instead, it followed the snow-covered chunks westward, deeper into Hay Bay. But after a couple of hundred yards, it faced bare ice again and retraced its tracks to where it had left the pack. Still it would not follow. It retreated back toward Hay Point and the shore, where the snow cover was continuous. Meanwhile, the pack had reached the far shore. The lone wolf took a longer route, staying closer to shore, and hurried to the pack.

With so much ice on Lake Superior, the pack certainly could have left for the mainland. Conversely, new wolves could have traveled to the island. But we never saw any sign that wolves left or that new wolves arrived.

As Don and I flew over the island that winter, we quickly noticed the wolves were grouping in numbers we had rarely seen before—thirteen wolves, ten, seven, five. We had been used to seeing the large pack of fifteen, at times with a loner tagging along behind. We were puzzled at first. Were these new wolves, different packs? We found these odd groupings traveling and feeding at kills together. The numbers were always changing. And then we would see fourteen or fifteen lying on the ice together.

We gradually figured out that these groups—at least most of them—were members of the large pack we had followed for the previous two winters. For whatever reason the animals were spending more time split into smaller groups. We didn't know why. We had suspected that the large pack probably split into smaller groups during the summer. Were the wolves simply getting a head start on this seasonal pattern this year? Were social interactions among pack members causing the pack to fragment? Or were there advantages to hunting in smaller groups, perhaps because of the deeper snow or because there appeared to be fewer moose calves this winter?

Certainly, as Adolph Murie noted after watching wolves in Alaska, the size of a wolf pack was "limited by the law of diminishing returns." And we saw how this might be true in our own research. In many chases and successful attacks, only a handful of wolves could really be in on the action. The rest were stragglers and bystanders. During the chase some fell far behind. Yet when the deed was done, they would all be there to divvy up the spoils. Under some conditions, it would be more efficient if a large pack split up into two packs to encounter and potentially kill twice as many moose. In fact, by the time our winter season was over, we had discovered that the large pack, often split into small groups, killed and ate more moose than in previous years.

As winter wore on, I began to suspect that a pair of wolves we often spotted were not part of a pack of three, as I had originally thought, but in fact formed a small pack of their own. That put the total number of wolves on the island this winter at twenty-one or twenty-two: the big pack of fifteen, a pack of three, a pack of two, Homer (the lone wolf), and perhaps one other loner. Often we saw only fourteen of the big pack at a time, and I suspected that one member occasionally peeled off to spend time with Homer. If this were true, I could only feel happy for him. We also realized that the large pack was spending nearly all of its time on the southwestern two-thirds of Isle Royale, leaving the northeastern end of the island to the smaller packs.

Nonetheless, the big pack continued to enforce its territorial prerogatives. In early March, the large pack was traveling alongshore near Cumberland Point when two other wolves dashed out of the woods, where they had been feeding on an old kill, and sprinted across the ice about 125 yards ahead of the big pack. The pack immediately took up the chase. One of the two wolves cut back into the woods, and the pack ignored it. The second and smaller of the two continued fleeing down the ice alongshore. It stopped briefly, looked at the pursuing pack, and assumed what I interpreted as a submissive gesture, stretching its front legs forward and lowering its head and shoulders—"down dog" for yoga practitioners. Apparently it didn't have much confidence in this tactic as the large pack bore down, and soon resumed its flight down the shoreline. The pack could not gain ground and eventually gave up, while the single wolf continued at full speed for another mile down the ice before dodging into the woods.

If hunting as a large pack allows wolves to kill large prey, what do pairs and lone wolves do to survive? We had often wondered this and saw plenty of evidence that single wolves scavenged kills made by the big pack. But we also saw lone wolves try to kill their own food.

Don and I were flying over the center of the island when we spotted a single wolf traveling upwind across some open ridges. It seemed to smell two moose about a quarter mile away. Following the scent, the wolf sneaked to within about 25 yards of one moose and then charged. Both moose jumped to their feet and took off. After several hundred yards, one moose stopped, and the wolf passed it by to follow the moose that was still running. But after several hundred yards more, the wolf gave up.

It wasn't finished yet. The loner continued up a slope covered in second-growth forest and scented another moose. It slowly stalked it. But at a distance of about 15 yards, the moose apparently decided to put an end to this game and strode boldly toward the wolf. The wolf cowered, thought better of the situation, circled around the moose, and continued on its way. The incident certainly shows the advantages to a wolf of hunting in a pack but also suggests that if a lone wolf was willing to engage a full-grown moose, the tactic must work at least once in a while.

We did watch once as a single wolf successfully attacked a full-grown moose, and it provided one of the most grisly events of my Isle Royale experience. Although I did not personally observe the most unpleasant part of the attack, ranger Ben Zerbey, who took a turn in the aircraft at that moment, told me about it.

We had been out of business for a few days as Don flew the Champ back to Eveleth for its hundred-hour check. When we got back in the air, we found the large pack at a new kill near Lake Halloran. We decided to backtrack the wolves to figure out where they had been earlier. As we followed the back trail, we noticed a lone wolf also backtracking the big pack. We soon understood why as we spotted a badly wounded moose lying on a hillside. It was surrounded by blood-soaked snow and wolf tracks, suggesting the large pack had attacked and injured it, probably several days earlier.

We circled and watched the lone wolf while it headed toward the moose. As it began to run excitedly along the trail, I had little doubt it knew the moose was there. About 50 yards away, the moose lay in the snow, and the wolf approached cautiously. As the wolf drew closer, the moose struggled to its feet. The wolf circled the moose at a distance of only 10 feet, and after a few minutes lay down nearby. A few minutes later, the moose lay down. Immediately, the wolf approached, forcing the moose to its feet. The wolf moved away and lay down. After a few minutes, so did the moose. Again, the wolf jumped toward the moose, wagging its tail. It seemed to snap at its quarry's nose but missed. The moose simply stood. The wolf lay down, but the moose stood for a half hour, when we had to leave to refuel.

We were back within an hour, the wolf sprawled on its side in the snow, the moose still standing. At this point, we flew back to check the whereabouts of the large pack, and then Don took me back to Washington Harbor. He returned with Zerbey, the chief ranger who had spelled Roy Stamey. As they flew over, the wolf was tugging and chewing at the moose's rump as the moose lay quietly with its head up, watching the wolf literally eat it alive.

Don returned Zerbey to camp and flew me back to the moose, but by then it was dead. I hiked in to the site the next day and discovered it was an old bull, about eleven years old, covered with winter ticks. Its lungs were packed with cysts from the hydatid tapeworm. The worms impact the lungs with golf ball–sized cysts that greatly reduce lung function. A rubbery yellow mucus clogged the lung passageways.

For more than a week, the lone wolf fed on the moose without competition, except for foxes and a few ravens. The large pack, at least as long as we were on the island, never returned there. They must have had enough fresh food from their new kills.

The poor moose that was partly eaten alive started at a severe disadvantage. It was old, diseased, and wounded. At that point, it was doomed, even against a lone wolf.

Healthy adult moose, by contrast, are a formidable foe, even when wounded. And as we learned observing many encounters this winter, moose are especially threatening if they stand and face down their attackers. Their kicks are swift and powerful and when the moose are standing still, can be delivered with great accuracy. When a moose held its ground, the wolves would try to force it to run. If it did not, they quickly abandoned it. We saw many examples.

Once we watched as seven wolves scented a moose in heavy blowdown near Lake Halloran and started toward it. The moose detected the wolves closing in from 20 yards away and might have made good use of the tangled trees to make its escape. Instead, it charged as the wolves closed in. The wolves scattered. They stood around for thirty seconds as if surprised and then called the whole thing off.

A few minutes later, the same wolves found three moose standing in a spruce thicket and the same heavy blowdown. The wolves ran to the first moose, which stood its ground. The wolves immediately turned their attention to the second moose, which began to run. With a head start through the heavy cover, it outpaced the wolves, which now turned to the third moose. This animal ran as well, feinting and attacking the wolves as they ran alongside. One wolf got ahead of the moose and seemed to present an inviting target for the enraged moose, which charged and chased it 50 yards down a trail, even with the rest of the pack nipping at its heels. Finally the moose sidled up to a spruce tree and confronted the wolves. In seconds, they turned and walked away.

As they say, the best defense is a good offense. A pack of seven or eight wolves scattered three moose near Hay Point. They ran off after one moose in a futile attempt to bring it down but soon gave up the chase. As they resumed their march along the shoreline, they scented one of the other moose in the woods about 150 yards away and began to close in. But this time, instead of running, the moose strode directly toward the oncoming wolves. For 70 yards, it kept coming, plowing through the snow toward the wolves. When it got to within 30 yards, the wolves had had enough. They turned and left.

We also saw a cow with a calf successfully face down wolves. As seven wolves traveled the Lake Superior shore near Lake Halloran, they veered inland toward the cow and calf resting in heavy cover. With the wolves only 10 yards away, the moose jumped up. The cow charged and then covered the rear of her calf. The two moose walked off a few steps. The wolves followed, trying to coax the moose to run, but they stood tight. After thirty seconds or so, the wolves returned to the shoreline, and the moose faded into the woods.

It was evident that moose that confronted wolves enjoyed full command of the situation. That begs the question, why would any moose choose to run? Cows with calves, perhaps, do better by running. And if a moose can get a jump on wolves, especially in deep snow or heavy blowdown, it, too, stands a good chance of escaping. Perhaps others simply lose their nerve: surrounded by a dozen or more snarling, snapping wolves, they panic. (And that may be an advantage of a large wolf pack—intimidation that forces a moose to flee.)

There might be something else going on as well: perhaps the moose that defy their attackers are confident for a reason. They might be healthy and strong. Those that are aware of their frailties might know better than to stand and fight. They flee and hope to get away with it, realizing that to confront the wolves would be their demise. That imputes a high degree of deliberation to a moose, but it may represent what they know innately.

Dr. Allen spent much of his time engaging the large pack, which spent more time than usual hanging around the Windigo ranger station on Washington Harbor. More than anything, he wanted good photographs of wolves.

It began as we hauled the carcass of our backdoor moose down to Washington Creek, where we had set up a lean-to for photography. One afternoon I helped him rig his remotely controlled camera next to the carcass in hopes of getting close-ups of wolves feeding at a kill.

A few evenings later, we heard howling from the head of Washington Harbor, not far from the cabin. It started at nine o'clock, well after dark, and lasted for two hours. The pack seemed to be on the ice of the harbor at first, and then they retreated up Washington Creek, deeper into the woods behind our quarters.

The next day, Dr. Allen spotted the big pack on the ice at the mouth of Washington Creek. He headed down to the lean-to to take photographs, while I watched from camp all morning. It was balmy—nearly thawing!—but a strong wind blew off the lake. The wolves eventually moved back up the creek, out of my view. But in early afternoon, at least fourteen wolves came running from the creek, down the harbor and a half mile out on the ice in front of camp. They ran quickly, as though they were scared. The only time I had seen them run so fast, when they weren't chasing a moose or a strange wolf, was when we had scared them. Perhaps they had scented Dr. Allen. I hiked down to the lean-to. Dr. Allen was still hale, hearty, and eager to get pictures. In fact, he remained on wolf watch until dinner. Afterwards he said the wolves had lain on the ice much of the morning. Though interested in the carcass, they didn't eat any of it.

The next morning, Dr. Allen was back at it—trying to photograph four wolves on the ice at the head of the harbor as I watched from the hill by camp. (Although he never got his prized close-up of wolves feeding, he did eventually manage to take some spectacular photos from the aircraft.)

At midmorning, a wolf howled twice from the woods and then appeared on the ice. It howled several times more—long, low, and drawn out, its muzzle pointed skyward each time. Soon another wolf appeared, then two more, and finally a fifth. They walked over the ice in the harbor and lay down. After a long rest, one lifted its head and howled a few times, arousing the wolf nearest it. The first wolf walked over, tail up and cocked forward, and playfully nosed the second wolf. The second wolf rolled over on its back and reached its paws up toward the other wolf. They played like this for some time before the other three wolves rousted themselves and the pack disappeared up Washington Creek.

One day when we couldn't fly because of heavy snow, Don Murray and Roy Stamey were ice fishing near Beaver Island when they spotted a wolf standing at the mouth of Washington Creek. It continued to watch them, even as they fired up the Weasel and rode back to the station.

The wolves continued to show up around the station. More so than at any other time during my fieldwork, the wolves seemed to be starting to accept our presence.

The wolf scats I picked up along the trails each summer gave an accurate picture of what the wolves had been eating, even when we couldn't watch them.

The sample was biased toward the wolves' diet in summer, which was the only time I collected scats. But the story the scats told meshed nicely with what we were seeing from the air during the winter. Moose hair, especially calf hair, was by far the most common component of the wolf scats. Beaver fur was significant but far less common.

We knew wolves occasionally ate beavers. We assumed the island's beavers, tucked into their lodges with underwater caches of green branches nearby, were safe from wolves during winter. But thaws in late February and early March opened up some of the beavers' waterways. The animals got to work early, eager apparently to gather fresh food. We spotted holes in the ice and beaver trails to trees, sometimes as far as 100 feet away from water. In the process, the beavers became vulnerable to wolves.

We found the first beaver kill near Washington Island, where a trail led from a hole in the ice near a dock and ended a few feet from shore in a big splotch of bloody snow tamped down with wolf tracks. From the plane, we spotted a wolf nearby, gnawing what easily might have been a beaver hide.

Later that day, we discovered the remains of a beaver near a small island in Tobin Harbor, at the opposite end of Isle Royale. Following wolf tracks, we deduced that the pack of three at that end of the island had investigated two beaver lodges located along the shore of the small island and had followed beaver trails to some freshly cut saplings. Along one trail was another splotch of bloody snow compacted by wolf tracks. Later we landed, and I collected a well-gnawed beaver skull, nearly cleaned of flesh.

While beaver fur showed up in the scats, the fur of snowshoe hares was scarce—less common than grass, only slightly more prevalent than soil. Which made sense. We knew wolves to be big-game predators, too large to efficiently scamper around and poke into alder swamps for hares.

The few encounters we saw from the plane confirmed that belief.

In early February we followed a lone wolf near the shore of Lake Desor. A single wolf, of course, would be at a grave disadvantage in tackling a moose. If any wolf was predisposed to chase hares, it would be a wolf that was alone. But we watched as one hare took off in front of the wolf, ran a circle around it, and seemed to take cover under a log. The wolf ignored it, even paying no attention as it crossed the hare's trail through the snow. Twice more we saw the wolf scare up hares as it walked through the woods. It didn't react at all.

A few days later near Lake Harvey we watched the large pack jump a hare. None reacted. Later in the day the same pack jumped two hares simultaneously and ignored them both.

To meet its need of 5 to 6 pounds of meat or more each day, a wolf would have to kill at least two hares. That would be a lot of work for a wolf, which is neither as agile as a fox or as able to penetrate blowdowns and alder thickets or to chase hares over snowdrifts without breaking through. These wolves had apparently decided, in their instinctive way, to concentrate on moose. (In some circumstances, however, a pack of wolves will chase snowshoe hares, as I later observed in Minnesota.)

We witnessed a peculiar and entertaining relationship between the island's wolves and ravens.

A flock of ravens often played leapfrog with the pack of fifteen as it traveled. The birds would fly ahead of the pack, perch in trees until the wolves passed, and then fly ahead of them again.

Sometimes I watched ravens trailing wolves, flying directly along a string of tracks, in an effort to find them. Discovering a fresh scat in the trail, a bird would land, pick the turd apart for edibles, and resume its pursuit.

Once when I watched the wolves attack a moose, the ravens swirled around in obvious anticipation. The wolves wounded the moose, and one bird sat in a tree and called as the wolves tried to force the moose to run.

At every new kill, ravens would flock in the surrounding trees. As soon as the wolves stepped away, the birds would swoop in to peck at the carcass. Sometimes the ravens, like the wolves, would eat bloody snow. Quite likely the wolves' leftovers were the ravens' primary winter diet.

One day, after the large pack had pursued an intruder wolf on the southwest end of the island, they crossed Grace Harbor and several wolves flopped down on the ice. Four or five ravens that had been following the pack swooped down to play. The ravens would chase the wolves, flying just above their heads. They would dive at a wolf's head or tail, and the wolf would duck and leap at a bird. I watched as a raven landed near a resting wolf, walked over, and pecked at its tail. The wolf jumped up and lunged at the raven. I also saw a wolf stalk a raven, which lifted off above the snow at the last fraction of a second and landed again, just a few feet away, and waited for the wolf to attack again. As the wolves assembled and began traveling again, the ravens flew ahead and waited in trees for the pack to catch up and pass them.

These efforts seemed more playful than hostile. I observed these encounters several times, yet I never saw a wolf actually touch a raven. Nor did I find any remains of ravens in the wolf scats I collected. Either the ravens were keenly aware of the wolves' abilities, or the wolves had no real intention of catching the birds. Perhaps both. Each creature seemed to enjoy the games.

The other species that follows wolves is the red fox. And for whatever reason, we saw many more foxes this year than we had at any time before. In one day alone we spotted ten from the air. Since we weren't studying foxes and had devised no real way to count them, we couldn't say much about their total numbers or even their relative abundance. But we surely were seeing a lot more of them. One day we spotted two on a tiny island on Lake Desor. Don set the plane down for a closer look, and I climbed out onto the ice. One ran off, but the other stretched and then curled up as if to sleep. I eventually crept to within about 50 yards, taking photos, until it too decided enough was enough.

From what I had seen the previous winters, foxes have much more to fear in their dealings with wolves than ravens do. Nonetheless, they often showed up at wolf kills. We spotted one fox feeding on a moose carcass while several wolves stood only about 150 yards away. Another day, a fox waited its turn as a wolf fed at a moose carcass only 100 yards away and another wolf stood a quarter mile away.

Despite the risk, the prospect of fresh moose meat was too good to pass up.

On March 20 (the first day of spring!) Don and I took off and quickly found the pack of fifteen wolves lying on the ice at the mouth of Washington Harbor, not far from camp. After taking photographs, we headed northeast along the length of the island and found a lone wolf—was it Homer?—feeding on a dead moose near Lake Desor. Continuing northeast, we followed the tracks of three wolves in Robinson Bay but soon lost the trail. After lunch, we checked on all the wolves again—the fifteen now near Rainbow Cove and the lone wolf still near the dead moose. We spotted the three wolves that had eluded us earlier on a beaver pond be-tween Lake Richie and Lake LeSage. On our way back to camp, we again photographed the large pack on Rainbow Cove.

And then it was over. There was still plenty of light (the days were much longer now), but we landed and spent the rest of the day packing in preparation for leaving the next day. My days of flying over the wolves of Isle Royale were finished. Three years of the most important and exhilarating fieldwork a young graduate student could imagine had come to an end. Oddly, I don't recall that I was wistful in any way—only that I was excited by what I had learned and eager to pore over my findings and complete my dissertation.

THE WOLVES OF ISLE ROYALE

ᘓᕢᕞ

Sometime after we wrapped up our winter season, Dr. Allen suggested that I head to Isle Royale for one more summer to teach Phil Shelton what I knew of working on the island.

Phil, one of Dr. Allen's grad students, was the newest member of the Isle Royale team. He was to focus his dissertation on the ecology of beavers on the island, but during the winter—not a productive time to be studying beavers—he would continue the aerial field surveys as part of his research assistantship at Purdue. I knew the island thoroughly, so it made perfect sense to help Phil. Dr. Allen knew that I would jump at the chance to go back for one more season.

So in early May, Betty Ann and our kids, Sharon and our new baby, Stephen, again settled into the Bangsund cabin on Rock Harbor. Betty Ann took care of the kids, continued her arts and crafts, and sold her paintings and wolf track ashtrays at park headquarters. We continued to visit with the island's summer residents, especially Pete and Laura Edisen.

We did have a bit of a family emergency, however. Steve, who had been born just days after Betty Ann had returned from Isle Royale last year, had had a high fever for a week. I finally headed to park headquarters and called a doctor on the mainland, who suggested we bring him in. So our whole family flew to Houghton, where Steve was examined and given medicine, and we returned to the island that afternoon.

We enjoyed quite a social season in what would be our last summer together on the island. My folks visited again, but with my maternal grandparents as well. Dr. and Mrs. Allen camped for a few days there, and friends of both Betty Ann's and mine stayed with us.

One of the folks who visited Isle Royale that summer was Fred Truslow, a wildlife photographer for *National Geographic*. Dr. Allen and I were working on our article about the Isle Royale wolves for the magazine, and Fred's assignment was to get some photos to supplement ours. Thus, I felt some kinship with what Truslow was doing. Moreover, with my continuing interest in writing and photographing for popular audiences, I made a point of asking him lots of questions, guiding him around the island, and even helping him carry a small dinghy into Lake Ojibway.

Meanwhile, I had been honing my own skills in reaching the public in other ways, speaking about the wolf project to the Isle Royale Natural History Association, YMCA groups, and Explorer Scouts who visited Rock Harbor Lodge.

Phil wouldn't be coming until mid-June. In the meantime, I continued my summer fieldwork, hiking trails, bagging scats, and making notes of the wildlife I saw. Snowshoe hares were still abundant this year: I saw fifteen during a single 3-mile hike one evening. There were plenty of sharp-tailed grouse, too. And my outboard performed as well as ever—sputtering, stalling, breaking, and necessitating several trips to see Frank Taddeucci, the park mechanic.

It was good to be home again.

I continued my search for the ever-elusive active wolf den. A few days after arriving, I hiked in to the old den we had found two years earlier along the north shore of Siskiwit Lake, in hopes it had been dug out and put to use again. As I neared the site, my heart leapt—it was occupied! Even from a distance I could see fresh dirt excavated from new tunnels. And there were fresh tracks in the soil. But up close, my hopes were dashed. The tracks had been made by foxes.

As I surveyed the tracks, I heard growling from the den. Then a nose poked out, and I could see a fox pup's head. I screwed a telephoto lens on my camera and gingerly waited about 10 feet away. One by one, six pups emerged from the den and huddled at the entrance. One climbed the mound and scratched itself. Each pup measured only a bit more than a foot long, excluding tail, and probably weighed a pound or two. They frolicked near the den, ducking in and out of the entrance, showing little concern for my presence. I waited, taking photos, for about ten minutes, but never spotted either of the parents. Just nine days later, I visited the site again, but by then it appeared to have been abandoned. No fresh tracks were anywhere to be found. The photos, however, would appear in articles, newspapers, a magazine cover, my book *The Wolves of Isle Royale*, and even a calendar. What a stroke of good luck!

I also had several encounters with moose that summer. Some were enlightening, others simply surprising.

Despite the many hours we spent airborne over Isle Royale in winter, we still knew very little about how cow moose might protect their calves from wolves in summer, mere weeks after their birth. So I was fascinated in late May when I startled a cow and calf along the Minong Ridge Trail. The calf was very small, perhaps only 25 pounds. It couldn't keep up as the cow moved away, so the cow was forced to wait. A calf that small wouldn't stand a chance in trying to flee from wolves unless the cow were to stand her ground to protect her vulnerable offspring.

Three weeks later, I got a taste of how a cow might react were I a wolf. I was photographing moose in Lake Ojibway when a cow and two calves came up the trail behind me. They were only about 30 feet away, and since there were no trees nearby to climb, I waved to alert them to my presence before they got any closer. The calves tucked in behind mom, who grunted and laid back her ears while I fumbled with my camera. Then she turned and walked slowly away as the twins followed close behind.

Les Robinette, a U.S. Fish and Wildlife Service biologist on the island, told me of a cow moose that made her defense even more emphatically. South of Siskiwit Lake, Robinette had come upon a cow and newborn calf, the remnants of its umbilical cord still evident. He climbed a tree to watch them. Each time he changed position, the cow snorted and raised her hackles. After about forty-five minutes, he clambered down to find another tree to get a better view. As he touched the ground, the cow charged. Robinette dove behind a tree as the moose straddled the trunk with her front legs and pounded the ground with her hooves. Their noses—Robinette's and the cow's—were only inches apart, a cloud of hot moose slobber and snot in Robinette's face. Having made her point, the cow then strode back to her calf.

One evening in June, I lugged my phonograph and wolf records to Lake Richie. It had been raining most of the day, but now it was clear and calm, and a hike offered some relief from reviewing and abstracting literature for my dissertation. I teed up the record and played a wolf howl. No reply, but immediately two crows flew in to see what might be in store. Across the lake, a moose snorted, crashed into the water, and swam to an island, grunting with each stroke.

A couple of weeks later, I got a much closer look at a moose (though not as frighteningly intimate as Robinette's encounter). Our family was eating dinner in our cabin. As always, I sat with my back to a small window, and Betty Ann across the table facing me. As we ate, Betty Ann's demeanor suddenly changed—her eyebrows raised, her eyes widened, her mouth opened, and she pointed toward me. "Look!" she gasped. I turned around, and there—2 feet behind me, nicely framed by the window—was the forlorn face of a cow moose staring in at us. Whatever she was thinking, we had no idea, but we had great fun speculating about whatever might be going through her mind.

Seeing my first moose three years ago, I had been struck by its bearing of dullness, of utter cluelessness. First impressions can be deceiving, as I discovered in my subsequent observations of moose, especially from an airplane. But it was true that moose were sometimes oblivious to dangers they had not evolved with. One afternoon, Paul Lamarour, the Mount Ojibway fire tower observer, stopped by the cabin to tell me that a moose was stuck in the Haytown Mine near the Todd Harbor campground on the north shore of the island.

The park organized a rescue party, and the next day we set out with several rangers, ropes, and block and tackle. The Haytown Mine was a steep-sided pit, excavated in bedrock by the copper miners who had canvassed the island a century earlier. When we arrived, we found a live moose at the bottom of the pit, about 30 feet down. How would we ever get him out? I had no idea, and the rangers were in charge, so I let them try to figure it out. No one volunteered to climb into the pit with the moose, which wouldn't have done any good anyway. Instead the rangers tied a noose in our five-eighths-inch nylon rope, dangled the rope over the edge of the pit, and eventually lassoed the moose—around the neck.

I had accompanied this team because I was always on the look-out for a moose I could necropsy, and it began to look like I would have an opportunity here. The rangers ran the rope through the block and tackle, and we all pulled for all we were worth, but the moose didn't budge. But it did nearly asphyxiate. So we tied the block and tackle to a pole lashed between two trees and tried again. We all pulled—me, four rangers, plus two adults and five boys from the campground that we recruited. One of the trees, a live balsam, 8 inches across, snapped at the base. We continued haul-ing and pulling, and suddenly—*pow!*—the rope snapped.

It had been clear that this attempt would only kill the moose, so I had positioned myself at the head of the line of pullers and only feigned pulling. Thus, when the rope snapped, I was the only one left standing as everyone behind me went flying backwards and fell over. The moose had been coming up the side of the pit, but it fell to the bottom and remained motionless. Some rescue.

Since I was the guy most interested in moose, it fell to me to clamber down the rope into the pit. The moose was indeed dead. When I opened its chest, its lungs were still blown up like huge balloons. I necropsied the unfortunate creature and quartered it. We brought all four legs back to the Mott Island mess hall, and everyone had a feast.

Once Phil Shelton arrived aboard the *Ranger III*, I set aside most of my fieldwork, which was virtually complete anyway, and began helping him with his beaver studies. We hiked the island, motored along its shores, and waded up marshy streams and swampy meadows looking for signs of active beaver colonies.

A necessary part of Phil's study was livetrapping beavers—to weigh and measure them, tag them, check them for parasites, and count the number of adults and kits in each colony. We used Hancock live traps, which looked like a large suitcase with a steel frame and chain-link mesh that clamped down around a beaver that tried to take food from the pan. We set them beside dams and lodges and along drag paths on shore, baiting them with whatever the beavers seemed to be eating at the time, including aspen, birch, red osier, willow, and bigleaf aster.

Trapping beavers was a gas. I loved the puzzle of finding suitable trap sites, the challenge of designing a good set, and the anticipation of returning to the traps to see if we had succeeded. It replicated the fur trapping I had done—the very thing that had gotten me into wildlife biology in the first place. Slopping around in beaver muck in shorts, we handled live animals—we even caught a 20-pound beaver on land by hand, grabbing it by the tail and holding it as you might a snapping turtle. It was up close and messy work, very different from the flyover observation I had done with wolves and moose.

Soon we were catching dozens of beavers for Phil's study. As is often the case with almost any kind of trap, eventually an accident happens, and an animal is caught wrong and dies. That is how I obtained some beaver meat to try.

As a young teenager trapping muskrats, I had learned that many types of wild meat were not only edible but delicious. This included muskrat as well as raccoons and possums (as long as you remove all the fat from both of these animals). My dad hunted, so our family regularly ate cottontail rabbits, deer, pheasants, and other critters. And as a bear researcher for several summers in my undergraduate days, I had feasted on meat from several black bears, which for certain parts of the study had to be euthanized. I had also read that beaver meat was relished by many outdoorsmen. The beaver meat, which I filleted primarily from the legs, was delicious.

I had also heard that some Europeans relish various kinds of brains. So I opened the skull, spread the brains out on a piece of aluminum foil, baked them in the oven, and then spread them on toast. I don't recall whether we salted and peppered them, and maybe we should have used a bit of ketchup, because it turned out the baked beaver brains on bread really could have used something. Anything.

I liked Phil a lot. I knew him from Purdue, before we worked together in the field. He loved classical music, as did Betty Ann and I. We had all become good friends. He was lanky and blond, long hair swept to the side. In the field he grew a faint beard and mustache. He was very bright—quicker than I was, I thought. On Isle Royale, after we had spent the morning processing and releasing a beaver, we went back to my cabin, where Phil translated a German paper on wolf social behavior for my dissertation.

But Phil was also an atheist, and I had grown up a Catholic. I remained devout in college, and when I had a family of my own, I made a point of attending church regularly. During my fieldwork on Isle Royale, I often attended mass at the little church at Rock Harbor Lodge. That was especially true once Betty Ann joined me.

As an undergraduate at Cornell, I had been focused entirely on getting through school—classes full-time and working thirty hours a week in a grocery store besides. But as a graduate student, especially as a graduate student with a family, I began to look at the wider world around me, beyond the horizons of wildlife biology. And that involved social things, not only going to church but reading the local newspaper and thinking of how I fit into the community. And politics.

It was a time of intense concern over Communism—*"godless Communism"*—especially in the church I attended. The church newspaper often railed about the Communists attacking the foundations of religious belief and the church. At Purdue, some professors were fervent anti-Communists and religiously quite conservative. As I looked for meaning and social involvement beyond the world of wolf biology, I too became a conservative activist and anti-Communist. In fact, during my last summer on Isle Royale, I mimeographed anti-Communist literature to hand out to people at the park and even organized a weekly anti-Communism study group. Thirty-two people attended the first meeting at Mott Island, and fifty-three attended a similar meeting at Rock Harbor Lodge.

So into this milieu walks Phil Shelton, the freethinker. Boy, did that make for some hard conversations on the island. But I liked Phil, and otherwise we got along well. I respected him. I could see that his viewpoint was rational. He could support his religious beliefs. That certainly wasn't true of many of the people who showed up at my meetings on Isle Royale or back at Purdue.

And there were also the John Birchers to contend with. This ultraconservative and hostile fringe was omnipresent in Indiana, where Purdue was located and where, in fact, the group had been founded. As I began to involve myself in this anti-Communism movement, I encountered racist statements and rants. I was taken aback. Fundamentalist religion and endemic racism were attitudes I had never encountered in upstate New York, where I grew up.

Not that it didn't exist, but I had never been aware of it. My dad was a union member in a chemical factory. That didn't exactly guarantee he was a social liberal, but I never heard him disparage people for their race. I say that because when I heard those kinds of comments from another acquaintance, they disturbed me, and I had words with him.

So the more I talked to Phil, the more I understood and valued his reasoning and point of view, and the more disillusioned I became with my fellow travelers.

During the rest of the summer and for quite some time afterward, I tried to disentangle my religion and my politics. The religious part was to come to a head a couple of years later as I met weekly with a Catholic priest to try to come to grips with my failing faith. "Can you save me?" I asked. But he couldn't, and after several meetings I realized that I, too, had become a nonbeliever.

My turmoil over politics and life philosophy lasted longer. I grew disenchanted—not only with my anti-Communist organizing but even with my wolf research. I began to consider a new career, one that had more to do with the human condition. The world was going to pot, I feared, and here I was counting wolves. That bothered me so much that I needed to do more than just study animals. I decided to finish my dissertation, obtain my PhD, and then pursue a degree that would allow me to study human-related fields such as anthropology, psychology, philosophy, logic—all fields I had missed at Cornell and Purdue. I was accepted into the doctorate program in American studies at the University of Minnesota, one of few such programs offered at the time.

This was the turmoil that enveloped me as I attempted to complete my dissertation on the wolves and moose of Isle Royale. I was not just conflicted, but now I had chosen a new path. Even so, the doctorate in wildlife biology was something I never had doubts about completing. I had come too far for that.

But I had also developed another problem, even more pressing in the short term—stress! As I studied and wrote, I was straining my eye muscles, causing pain that interfered with my writing. They wouldn't relax, even after a full night's sleep. I ended up taking a train from West Lafayette, Indiana, to Chicago for treatment every week to get my eye muscles to relax. In addition, Dr. Allen arranged for a special project livetrapping cottontail rabbits in the fields near Purdue to keep me funded as my eyes healed and I continued to write. Finally, by the spring of 1962, I had finished my dissertation.

My disenchantment aside, Dr. Allen was enthusiastic with the results of the fieldwork, and the project had received a lot of acclaim even before the dissertation was finished, in the form of newspaper stories, public-speaking engagements, and Dr. Allen's and my upcoming article for *National Geographic*. Shortly after the dissertation was complete, I revised it a bit for publication by the National Park Service as *The Wolves of Isle Royale*. It was the seventh monograph in the National Parks Fauna series, and as Dr. Allen wrote in the introduction, it laid "the groundwork for the continuing program that is necessary to gather significant information on such long-lived animals as wolves and moose. . . . This account of the great wild dog of North America and its largest antlered prey has something important to add."

By far my most important finding was this: Wolves, even my big pack of fifteen, fail in their efforts to kill a moose far more often than they succeed. Once I had figured out I could watch these wolves hunt, I began to realize how often they failed or even neglected to take up the chase. The wolves detected many moose but confronted or chased barely more than half. Of those they pursued, the wolves killed fewer than 8 percent. Wolves have to struggle to earn a living. Adolph Murie, in Alaska, had observed that wolves there did not always succeed while hunting Dall sheep and that they killed mostly vulnerable ones. However, because he could not follow the wolves regularly, he couldn't estimate a failure rate. My learning specifically how seldom wolves succeeded at killing moose not only supported Murie's findings but began to make us realize that failure at hunting was probably a general characteristic of the wolf's life. As killing machines, wolves are not that well oiled.

The second insight was closely related to the first: The wolves spent much time and energy "testing" animals to find targets that were vulnerable either because of their size, health, age, or perhaps unfortunate circumstances (such as the moose that plunged over a low cliff, fell into a snowdrift, and was immediately swarmed by pursuing wolves). I recognized how all the apparent failures fit with Murie's finding that wolves kill primarily the weak, old, and sick. Each failure was in a sense a test to find the animals the wolves might succeed in killing.

Once I amassed and analyzed enough moose mandibles and observations to see which animals the wolves were killing, I realized the wolves rarely killed healthy animals in the prime of life. Clearly even a large pack of wolves struggled to subdue a healthy moose, whether it stood its ground or ran. I recovered and analyzed more than one hundred moose remains that I could "age" by tooth wear. Only eight were an estimated one to seven years old. All the rest were calves or old adults. As I later learned, wolves struggle to kill even smaller animals, such as white-tailed deer and caribou, if they are healthy and in the prime of life.

Finally, flying over Isle Royale for three winters and collecting scats during summer reinforced the belief that wolves are big-game predators that probably can't sustain themselves expending energy on small prey. Isle Royale wolves overwhelmingly ate moose. Beavers were clearly secondary prey. Animals such as hares hardly figured in their diet at all. Might wolves in a different setting, with different prey species to choose from, focus more on small species? Perhaps out of desperation. But given how dangerous and formidable moose are and how much difficulty wolves have in killing them, it was hard to believe they would completely ignore the opportunity to chase and kill hares if such small animals were a viable option.

Some of our findings were particular to Isle Royale. Dr. Allen had told me to count moose and count wolves. We did that, counting twenty-one or twenty-two wolves each year. We weren't able to say why wolf numbers remained so constant. Wolves can multiply quickly, but in this case their numbers hadn't grown. So that number probably represented something close to the carrying capacity of the island, given the number of moose and the condition of the browse that sustained the moose. But by what mechanism were the numbers so stable? Perhaps some wolves died or migrated off the island, and we never realized it because individual wolves were virtually impossible to distinguish from the air. Perhaps the large pack consisted of infertile or otherwise unproductive animals. Perhaps in a population near carrying capacity, wolves underwent some kind of stress-induced birth control. Were many unseen pups born? Did many die? We couldn't tell.

The moose were harder to count than the wolves. They probably numbered at least six hundred while I was on the island. From what we could estimate of calf production and the number of moose (calves and adults) killed by wolves each year, it seemed predators and prey were in some rough equilibrium. Of course, that was just during three years, not long enough to know if such stasis would persist. Still, I felt confident that both wolves and moose would continue to survive on the island for decades to come.

The presence of wolves seemed to have prevented the wild fluctuation in moose numbers that had occurred in the early twentieth century, before wolves appeared on the island. But again, three years was too short a time to know if that pattern would hold.

I also came to understand the routes the wolves followed around the island, the home ranges of the various packs, how far and frequently wolves traveled, and how often they killed and ate. And I learned that under these circumstances—wild wolves in a wilderness environment—wolves posed little danger to humans.

Finally, our project proved that Dr. Allen's instincts were right: Isle Royale was an ideal outdoor laboratory, simple and confined enough to study easily but wild enough to provide broad insights into wolf behavior. I think the project worked out better than either he or I had hoped.

The Isle Royale study positioned me to embark on a career I couldn't have imagined as a young graduate student. Of course, that was hard to appreciate at the time. After all, I was preparing to ditch wildlife biology, get a new degree, and launch an entirely different career. After I finished my dissertation, I put the family, including our new baby, Nicholas, born in April 1962, in the back of Betty Ann's Nash Rambler and drove to the Twin Cities, where I began my doctorate in American studies.

Fortunately, in retrospect, financial necessity resolved my vocational turmoil after a very enlightening year in the American studies program. With three children under five years of age, Betty Ann couldn't work, and I couldn't support the family on my half-time University of Minnesota graduate appointment. Disillusionment with the politics of the time didn't help either. I had to make a choice: we could starve for three more years as I strived to attain credentials for a new career, or I could instantly double my salary and lead a fascinating life as a wildlife biologist.

One day I left my office in Folwell Hall and walked around the block to the James Ford Bell Museum of Natural History, which was just starting the nation's first research program in radio-tracking wildlife. I forsook my American studies program and accepted a postdoctoral position working on the radio-tracking research.

A few years later, between jobs, I wrote a book for a popular audience called *The Wolf: The Ecology and Behavior of an Endangered Species*. It was a comprehensive look at the animal, based on what science knew at the time. And much of the science was mine, developed during my three-plus years on Isle Royale. I was lucky: the timing was good, the book became influential and popular, and now, fifty years later, it remains in print.

With that, my career finally was launched.

THE VIEW FROM AFAR

✿

After completing my work at Isle Royale, I stayed in touch with Dr. Allen and heard about the folks who carried on the study after me. My friend Phil Shelton continued his beaver research during the next couple of summers while conducting moose and wolf surveys in winter. Peter Jordan worked on the island for three years, completing his postdoctoral work on moose. Wendel Johnson came to Purdue from Michigan State University and studied the smaller mammals of Isle Royale. Michael Wolfe, who completed his PhD in West Germany before joining the Purdue team in 1967, worked on the island for several years, studying both wolves and moose.

Rolf Peterson joined the Isle Royale crew in the summer of 1970. As a kid Rolf had read Dr. Allen's and my article in *National Geographic* and stories of our work in the *Minneapolis Tribune*. In college he wrote to Purdue to apply to work on Isle Royale as a graduate student. He received a packet of boilerplate college materials and figured he had been brushed off. Only later, as he watched a television program on the project, did he recognize that he had written the wrong department. He contacted Dr. Allen, who invited him to apply. The next summer he was on Isle Royale. While I was Dr. Allen's first graduate student, Rolf would be his last.

Rolf was the right person at the right time. Dr. Allen was preparing to retire and was looking to hand off the project. Rolf had the interest and tenacity and soon acquired the outdoor and administrative skills to run the research. After finishing his PhD, Rolf published it in the same government series as I had, as *Wolf*

Ecology and Prey Relationships on Isle Royale. He then took the reins in 1976 as Dr. Allen retired to become a professor emeritus. Rolf secured a position at Michigan Technological University and moved the study to that institution. Even after retiring as a professor in 2006 and handing off the project to his student John Vucetich, Rolf has remained deeply involved in the studies, as has his wife, Carolyn. Together they raised their family on the island.

Rolf really expanded the Isle Royale project by introducing new tools and analytic techniques to the study. He started collaring wolves with radios. Some of this was inevitable with the march of time and technology and didn't require any real leap of innovation. But Rolf had the imagination and ability to recruit specialists who added new dimensions to the study. As technology developed to sequence DNA, he began to delve into the genetics of the island's wolves. He also teamed up with arthritis specialists to study that affliction in moose. Some of his findings may even have application to human health. Then Vucetich brought math and statistical analyses and rigor to the study. Rolf's work gave the Isle Royale research many new dimensions.

Furthermore, Rolf's longevity and continuity on the project from decade to decade provided perspective that might have been difficult for a rotating cast of researchers to attain. As specimens—including moose mandibles—built up over the years, he was able to reconstruct the population as it appeared through time, something that I, in the nascent years of the study, could never have hoped to do.

One of my favorite people from the study, pilot Don Murray, evolved from wolf hater to an avid admirer of the animal and the science being conducted on the island. He continued to fly the winter surveys for twenty years. I last saw Don in 2009, on Isle Royale, at the celebration of the fiftieth anniversary of the study. He died on November 14, 2014. Don's grandson, also named Don Murray, has now taken over the flying.

I stayed in touch with Dr. Allen. I visited him in Indiana in 1994, when he signed and gave me a copy of Stanley Young and Edward Goldman's *The Wolves of North America*, bound in bison leather and personalized to Allen by author Young. I ran into him again a few years later when I visited Purdue to receive an award. I was crushed in October 1997, when I learned that he had died of a heart attack. I really liked the guy, and of course I owed much of my success to his guidance and belief in my abilities.

A nationally renowned wildlife biologist, Dr. Allen was probably best known for having conceived, established, and administered the Isle Royale project, the longest-running scientific study of predator–prey relationships. Yet ironically, he wasn't really known as a scientist. He had his name on a few studies, but his real accomplishments lay in his writing, largely for a public audience, such as his book *Our Wildlife Legacy*. He was working for the U.S. Fish and Wildlife Service in Washington, D.C., when he first learned that wolves had reached Isle Royale. He was insightful enough to see it as a major research project. It was his idea. He laid the groundwork. He found the researchers. He nurtured the program and defended its integrity when the project was threatened by bureaucratic maneuvering. We owe much of what we have learned about wolves to Durward Allen.

I concluded my three years on Isle Royale in the winter of 1961 with, perhaps, an illusion of wolf–moose system stability. During my time on the island, the number of wolves had remained at twenty-one or twenty-two. The number of moose hovered at about six hundred. As I wrote in my thesis, "Apparently the Isle Royale wolf and moose populations have reached a state of dynamic equilibrium. Each is relatively stable, so any substantial fluctuation in one probably would be absorbed by the other until another equilibrium is reached."

If only I knew then what I know now.

After I left the island, the moose population slowly rose to 1,200 or so by 1972. Then it dropped for several years, until 1981. Again it grew, slowly at first but then rapidly, shooting up to 2,400 by 1995. It crashed to 500 two years later. It rose slightly and then dropped to under 500 again by 2006. During the next few years, it rose again to more than 1,000.

Meanwhile, what of the wolves? For a while, the wolf population grew as the moose population grew. But as the moose population crashed, the wolf population shot up to fifty in 1980. And then it crashed, dropping to fourteen in 1982, even as the population of moose slowly grew. During the early 1990s, as the moose population skyrocketed, the number of wolves grew only slowly. While the moose population crashed, the wolf population continued to increase slowly.

What was happening? We were discovering that the populations of wolves and moose were far more variable than we would have known from just my first three years of data. The relationship between moose abundance and wolf abundance turned out to be quite complex. Clearly other forces were at work. At the very least, these other factors complicated the picture.

The most obvious and perhaps profound influence—at least in hindsight—was weather. During my three winters, snowfall was light to moderate. But several times in the years since, the snow drifted to the height of the eaves of the ranger station, and researchers shoveled trenches to get around camp. Moose, with their long legs, are built to navigate snow. But when snow gets that deep, the burden of traveling to find food is enormous, and moose struggle to maintain their body fat until spring. Deep and prolonged snow can have a profound influence on the moose population. With moose in weakened condition, wolves cash in and can drive their numbers down. Even when wolf numbers and the predation rate are low, some moose starve. When snowfall over a few winters is light, moose gain better condition, and wolves have less effect.

Another factor we couldn't anticipate was climate change. With warmer winters, winter ticks have become more numerous. I found some moose with more than a dozen ticks on a single square inch of various parts of their bodies. It was never clear whether the ticks could actually kill a moose, but no doubt they weakened them. As winters have grown progressively warmer, ticks appear to be on the increase and presumably so is the stress on the moose.

In 1988 researchers discovered the antibodies of canine parvovi-

rus, a contagious and usually fatal viral illness, in the blood sam-
ples of two wolves, suggesting the disease had reached the island,
perhaps by way of an infected dog brought by a visitor. Rolf and
I disagree whether parvo could be implicated in the crash of wolf
numbers between 1980 and 1982. I believe the cause was starvation and wolves killing other wolves in conflicts over territory. (It's
not that unusual for scientists to disagree, but the disagreements
are professional, not personal.) The virus *could* cause the death
of pups and stymie the recovery of the wolf population. That it
appeared on Isle Royale suggested the outdoor laboratory of the
island wasn't as isolated as we had assumed.

From the standpoint of a predator–prey study, Isle Royale was
showing us that the relationship between moose abundance and
wolf abundance was complicated and differed over the years. Predation by wolves on moose did not always prevent the moose herd
from increasing. And the abundance of moose or their decline was
not always related solely to the number of wolves. It's not even
certain whether the existence of wolves on the island moderates
the ups and downs in moose numbers or drives them to greater extremes. What does seem important is that snowfall plays a critical
role in the nature of wolf–prey relationships.

Clearly, there was no smoothly operating "balance of nature."
Rolf and his team have asked, is nature as expressed on Isle Royale
the result of predictable patterns that researchers can discover and
use to predict the future? Or is ecology the process of "innumerable contingencies" that cannot be predicted but only explained
in hindsight? As Rolf has written, "Is ecological science like the
study of physics or more like the study of human history?"

While new data and insights poured in from Isle Royale, I was
studying wolves in other areas.

In northern Minnesota's Superior National Forest I flew over
wolves in wintertime, tracking their movements by radio collars,
replicating the Isle Royale work but over a larger area with more species of predators and prey. Here wolves shared their economy with
coyotes, bobcats, lynxes, and bears. Depending on the geographic

163

area, some wolf packs lived mainly on a "moose economy," while most depended primarily on white-tailed deer. Dr. Michael E. Nelson recorded radio-collared deer survival, movements, and interactions with radio-collared wolves from 1974 through 2010. Dr. Shannon Barber-Meyer is now continuing the fieldwork that we both did. This project has become the second-longest predator–prey study, after Isle Royale.

For nine years in Alaska's Denali National Park, my assistants, colleagues, and I tracked radio-collared wolves and caribou and their calves, studying patterns of prey selection by wolves, and causes of mortality in caribou. There, alternative prey were moose and Dall sheep, and fellow predators were black and grizzly bears.

In Yellowstone National Park, I helped reintroduce wolves in 1995 and 1996. My former PhD student Steve Fritts was in charge of the capture operation in Canada, and I was his consultant. The reintroduction effort took twenty years of planning and proved to be an immediate success once we released the first animals. Since then, through a series of seven grad students, I have continued to do research on the Yellowstone wolves and their prey.

On Canada's Ellesmere Island in the high Arctic, I finally was able to do what Adolph Murie had done—study wolves at their den sites. By habituating these wolves, which had never been hunted or harassed, to my presence, I was able to sit near the den and watch the care and feeding of the pups. The wolves would occasionally approach to within a few feet—close enough that I could easily identify individual animals and begin to recognize their habits and personalities. On an all-terrain vehicle, I could follow the pack for miles across the open tundra to watch them hunt musk oxen and arctic hares (three times the size of snowshoe hares). Research in open terrain, such as in Yellowstone and Ellesmere, has added immeasurably to our understanding of wolf behavior.

The years of study confirmed some of what I discovered about wolves on Isle Royale, refuted other things I "knew," and added substantially to our understanding of the animals. Of course, many of these things weren't "my" findings but simply corroborated what other field researchers were discovering. Decade by decade, there were far more researchers studying wolves and predator–prey interactions than there had been when I began my career.

In all these locations—all in northern climes with potentially hard winters—weather is the card that trumps all else in determining whether animals live or die. Whether it is deep snow in northern Minnesota, or the frozen armor of ice and snow that sometimes encrusts the Arctic tundra, extreme winter weather obliterates whatever "balance" may have existed between populations of predators and prey.

Bouts of starvation reverberate through successive generations—a "mother effect" and even a "grandmother effect" in which the young are weaker than they would have been had their mother or grandmother experienced good nutrition. A series of hard winters really weakens the prey population for a long time and increases its vulnerability to wolves. And a series of mild winters strengthens it.

Like the wolves of Isle Royale, wolves elsewhere struggled to kill their prey. Though I didn't always work in circumstances that allowed me to count and document success and failure, it was clear that wolves often "tested" animals and failed to kill far more often than they succeeded. As was true on Isle Royale, the wolves most often killed the young, the very old, and the unhealthy.

My grad student Dan MacNulty discovered in Yellowstone that hunting effectiveness plateaued as packs got larger and more wolves became "free riders," letting others (usually their parents and older sibs) do the dangerous job of killing big prey. On Isle Royale I imagined such a dynamic might be in play, but in Yellowstone we demonstrated it. Efficiency in hunting for elk seems to peak in packs of four to five wolves. With hunting bison, packs as large as thirteen or fourteen are most effective. Rolf Peterson and his colleagues have found that on Isle Royale, larger packs are also better able to fend off the scavenging of ravens, which can consume up to a third of what the wolves kill.

Contrary to what I and other wolf biologists propagated to the public early on, the concept of "alpha wolves" obscures more than it clarifies. That view of "top dog" suggests that wolf packs are collections of animals in a continual test for leadership and breeding rights. That may be true in groups of captive wolves, where this idea originated. But in the wild, as I discovered in many summers of watching wolves on Ellesmere Island, packs are simply families—a breeding pair and a couple of generations of offspring. There's dominance, sure, but it's the dominance of parents over children. To become an "alpha," all a wolf has to do is mate with another and start its own pack, just as with humans starting their families. Scientists no longer apply the term to wolves.

Wolves sometimes rely on small prey animals more than we believed, whether beavers in northern Minnesota or arctic hares on Ellesmere Island. But still, they do not live long without a staple of large hooved animals. Contrary to what Farley Mowat described in *Never Cry Wolf*, they cannot survive only on mice. (By the way, I found the book a great read—as fiction.)

As I started my research on Isle Royale, I imagined wolves to be a symbol of wilderness. So did everyone else. We believed that with a finite acreage of wilderness, wolf numbers, too, were finite. And that as we chipped away at what little wilderness was left in America, America's wolves would fade to extinction. A colleague in nearby Wisconsin, Dick Thiel, even determined a threshold of road density, above which wolves would struggle to survive.

We didn't fully appreciate that wolves really don't need wilderness so much as protection from human persecution. If they get that, through public acceptance or strictly enforced laws, they can survive in a landscape of farms, settlements, and roads. In fact, they can thrive.

With the protection of wolves under the new federal Endangered Species Act in 1973, I worked with several other wolf biologists on the *Recovery Plan for the Eastern Timber Wolf* in Minnesota, Wisconsin, and Michigan.

We had a long road ahead of us. Minnesota had fewer than 1,000 wolves, virtually all in Superior National Forest, including the Boundary Waters Canoe Area Wilderness. Wisconsin and Michigan at the time had almost no wolves at all.

By the time we finished the plan in 1978, wolves were already increasing in Minnesota. Soon dispersing wolves spilled over into Wisconsin and Michigan's Upper Peninsula. By 1999, the numbers exceeded the recovery goals of at least 1,250 wolves in Minnesota and 100 wolves in the other two states combined, for at least five consecutive years.

In 1987 the *Northern Rocky Mountain Wolf Recovery Plan* (for which I was only a consultant) proposed restoring wolves to Wyoming, Montana, and Idaho. Greater protections in Canada were already allowing dispersers to move southward into northern Montana. The passage of the Endangered Species Act had given impetus to the long-discussed idea of restoring wolves to Yellowstone National Park. As I mentioned, wolves were released in the park in 1995 and 1996. By 2002, the population had reached recovery goals of at least ten breeding pairs in each of the three state recovery areas for three consecutive years.

Wolf recovery in the United States has exceeded all our expectations. Thanks to Endangered Species Act protection and recovery programs, at least 6,000 gray wolves now live in the Lower 48. Their numbers are increasing, and they continue to move into new areas. Breeding packs now also exist in Washington, Oregon, and California. Individual dispersing wolves have been spotted in Utah, Colorado, North Dakota, South Dakota, Missouri, Indiana, Illinois, Iowa, Kentucky, and New York. The reintroduced wolf population in Arizona and New Mexico now numbers more than 130 and increases every year.

Despite wolf populations having long ago met the goals laid out in the recovery plans for the Great Lakes and Northern Rockies, only wolves in Montana, Idaho, and parts of Washington, Oregon, and Utah have been taken off the federal endangered species list, through direct intervention by Congress. Elsewhere, except Minnesota, where they are classified as threatened and subject to government depredation control, wolves remain fully protected by the Endangered Species Act, despite repeated attempts by the U.S. Fish and Wildlife Service to remove them from the list. Each time, the proposed delisting has been blocked in court by protectionist groups arguing on legal technicalities.

This continued full protection for flourishing wolf populations creates a real problem, not only for farmers and ranchers, who are generally hostile to wolves, but also for the wolves themselves. Because when wolves proliferate where they conflict with people, they will be killed—legally or not.

As we developed the recovery plan for Great Lakes wolves, we recognized that the rural residents who actually live close to wolves have a limited appetite for the animals. So we recommended that states such as Minnesota be zoned to allow lethal control of wolves in areas where conflict with humans was inevitable. We also envisioned that once recovery goals had been met, the wolf would be delisted and, where necessary, their numbers controlled, like those of bears, deer, geese, and other wildlife.

As wolf populations in the Northern Rockies and Great Lakes continue to increase far above recovery goals, animosity grows among many rural residents, especially ranchers, farmers, hunters, and hunting guides, who feel wolves are in direct conflict with them. These folks support lethal wolf control in far greater numbers than people who don't live in wolf country. That animosity undercuts the overall recovery efforts for the wolves. It leads to poaching and similar lawlessness, and increases hostility toward the Endangered Species Act.

When wolves are delisted, management will be returned to the states. The states all have management plans that would protect a sustainable number of wolves while dealing with problem animals or an overabundance through some form of regulated taking by professional wildlife agents, sport hunters, or landowners.

Sport killing has stirred outrage when hunters have shot wolves on the border of sanctuaries such as Yellowstone National Park, where individual wolves are often well known to wolf watchers.

In such situations, it would be possible to set up no-hunting buffers around parks where wolves are viewed by the public. Sanctuary zones elsewhere might also ease tensions with Indian tribes, some of which revere wolves and oppose their killing.

Opponents of wolf control argue that wolves should continue to be totally protected and allowed to proliferate throughout their current range until they disperse and establish breeding populations in all areas of the country suitable for them. That puts a tremendous burden on rural residents where wolves are already plentiful. I believe that even if wolves were controlled under sound state management, there would be wolves enough to disperse to the few new territories still available where wolves will not conflict too much.

We have emerged from a time when most people have held erroneous ideas about the wolf as a destructive and worthless creature. Now we are in a period when many people hold an opposite idea that is equally erroneous: that wolves are sacrosanct creatures that can magically restore "ecosystem health." Not just pro-wolf activists, but even some scientists are in danger of sanctifying the wolf.

In the years after wolves returned to Yellowstone, several scientific papers appeared that claimed that wolves had created a "landscape of fear" that disturbed the herbivorous habits of elk, allowing the lush regrowth of vegetation. Some of these studies claimed that this regrowth supposedly fostered a "trophic cascade" that reverberated throughout the ecosystem, affecting beavers, smaller predators, songbirds, and even the course of the rivers themselves.

However, closer examination and additional research failed to show such a clear effect. Wolf-mediated trophic cascades may exist—in fact, we expect them to. But any effect is too tangled up in all sorts of other complications and confounding factors to show definitively that all such changes were caused by wolves. For example, a recent article documented that cougars, more than wolves, tend to affect Yellowstone elk habits. As I wrote a few years ago, even if such cascading effects held true in national parks, they would have little relevance to most of wolf range because of overriding human influences on wolves, prey, vegetation, and other parts of the food web.

The wolf is neither a saint nor a sinner except to those who want to make it so.

Ironically, as wolves proliferated throughout much of their range and recolonized several new states, they were faltering on Isle Royale.

For several years during the 1990s, the population of wolves seemed stuck at about a dozen. Rolf and his colleagues were able to measure a remarkable increase in inbreeding depression, the loss of fitness that can occur from many generations of breeding within a tiny gene pool. Some wolves had extra vertebrae; the backbones of others were greatly malformed. The population was stagnating.

However, in 1997 a wolf later dubbed Old Grey Guy crossed the ice from Canada and began breeding with the wolves on the island. The influence of Old Grey Guy was immediate and profound. The population grew rapidly to thirty. Through DNA analysis Rolf's group was able to decipher what was going on, but that wasn't until several years later, and it showed that all the wolves on the island were now related to Old Grey Guy.

When Dr. Allen and I were beginning the study, we gave little thought to inbreeding depression, partly because the population was still new. Besides we had no way to detect or measure it. It was only decades later that the new field of molecular genetics made it possible to test for the lack of genetic diversity and presence of deleterious genes in a group of animals.

But Old Grey Guy's success eventually created its own inbreeding depression as the big male wolf mated with his own offspring, eventually siring twenty-one pups with his daughter. Soon the offspring were pairing up and mating. As Old Grey Guy's genes dominated the entire population, genetic diversity again declined. To make matters worse, three wolves, including a breeding pair, had broken through the ice and drowned in a historic mine pit. By 2010 or so, the population began trailing off again. By 2016, the entire population on Isle Royale had dwindled to only two wolves, which showed no signs of successful reproduction. Meanwhile, the moose population, numbering only 400 or so in 2006, had jumped to 1,600.

All this was very alarming—to wolf lovers, to devotees of Isle Royale National Park, and I'm sure to Rolf Peterson and his team, who were losing their research subjects, not to mention the whole basis of the predator–prey study. As the wolf population struggled and then foundered (even before the arrival of Old Grey Guy), Rolf sounded the alarm and argued that the population be "rescued" by bringing in new wolf blood from the mainland.

Rolf and John Vucetich argued that global warming was making the possibility of ice bridges and the chances of wolves crossing from the mainland less and less likely. Thus, if humans created this situation by burning fossil fuels, they had the right, even the obligation, to fix it, even in a national park. But what if, as I had read, climate change would increase the extremes of climate. That might produce even more frequent ice bridges. (And in fact, that is what has been happening lately.)

At first I was skeptical about interfering. This was a national park, after all, much of it designated wilderness, where policymakers and park advocates had long argued that nature should take its course. Presumably this sort of genetic tragedy happens all the time with island or isolated populations. We just happened to be studying this one intensively. By reintroducing wolves now, would the National Park Service be committing itself to intervening every time a population seemed headed for trouble? To quote Roderick Nash, an environmental historian who wrote the book *Wilderness and the American Mind*, would the Park Service choose to be "guardians" or "gardeners"? Marvin Roberson, the Michigan chapter of the Sierra Club conservation representative, suggested that a "cycle of manipulation" would eliminate the opportunity for a "novel ecosystem" to emerge naturally.

I felt similarly. At the very least, we should wait until the wolf population on the island had winked out before considering something as momentous as translocating wolves from the mainland.

After all, there were plenty of intriguing scientific questions that might be answered by taking a wait-and-see approach. Might the ship right itself? Might the wolves struggle through the crisis, or might another infusion of new blood from the mainland rescue the population? Was inbreeding the only culprit, or was it only one of many factors affecting the wolf population?

One of the most important lessons of Isle Royale has been the documentation that a tiny, isolated population of wolves could persist for decades and seem to function normally, despite high levels of inbreeding. This information was invaluable to wolf conservation and management, indeed to the entire field of conservation genetics. Shouldn't we allow the situation to play out before hitting reset?

In fact, had we intervened at the first signs of trouble, we would have missed the revealing effect of Old Grey Guy on the island's wolf population. That was an interesting story, and we wouldn't have seen it.

As I wrote in 2013, "In the medical field, when a threatening condition is detected that is not immediately causing distress, physicians often counsel 'watchful waiting'. . . . [T]he precautionary principle would weigh heavily in favor of non-intervention because once intervention is imposed, that condition can never be undone, whereas non-intervention can always be countered by intervention."

As it became apparent that the Isle Royale wolves were unable to reproduce, and the population was circling the drain, I had a change of heart. It was obvious the population would not recover and grow. And the political conditions for reintroduction looked favorable. I began to see the possibility of reintroducing new wolves as "An Unparalleled Opportunity for an Important Ecological Study," as we titled a paper I cowrote in 2017.

Fundamentally, the reintroduction of wolves was the only way to ensure that the island could continue to host a wolf–moose study, albeit a new one. We (that is, scientists) now had the opportunity to pick and choose the wolves to introduce. We might select a single pair and recapitulate the conditions of the wolves reaching the island in 1949. Or we might introduce as many as thirty unrelated wolves and see how they sorted out territories. Or we might introduce a few now and slowly add more later. We could document their genetics, parasites, and physical dimensions and see how those might change over time. We could make sure they started out with radio or GPS collars so we could see how they dispersed and colonized the island. We would be able to see how old wise wolves fared in an unfamiliar territory compared with younger wolves. We could study wolves familiar with white-tailed deer or caribou to see how quickly they adapted to hunting moose.

It really could be an unparalleled opportunity.

Of course, it wasn't up to me anyway. The National Park Service, after several years of consideration, approved in spring 2018 a plan to reintroduce twenty to thirty wolves over three years.

During that fall and winter, a team of biologists began capturing wolves and airlifting them to the island. Some came from the Grand Portage Reservation in northeastern Minnesota. Several were from Michipicoten Island in eastern Lake Superior. Others originated from the Canadian mainland along eastern Lake Superior near Wawa. By spring 2019, Isle Royale's wolf population had jumped from two to fifteen, and scientists were tracking their movements over the island.

There were a few mishaps and surprises. A single female from Grand Portage decamped across the ice back to the mainland. One female died in captivity before she could be transported and released. A black-coated male was found dead in a swamp on the southwestern end of Isle Royale several weeks after release.

During the winter as all this was unfolding, unexpected tracks showed up on the island. They appeared to be of different sizes, suggesting that at least two wolves had traveled over the ice from the mainland. Later the same day, field-workers found three sets of tracks along the north shore of the island that didn't correspond

173

with the location of any of the wolves known to be there. Researchers' best guess is that several wolves crossed the ice from the mainland, only to return there. Whether any remained behind or had the opportunity to breed on the island might become apparent only through DNA analysis of wolf scats or of any pups born on the island. It was ironic, perhaps, that after all this discussion of rescuing the Isle Royale wolves, mainland wolves had chosen this moment to make an appearance.

By early March 2020, twelve to fourteen wolves survived on the island in four groups, two of which were mated pairs occupying territories. Another mated pair and a bonded pair plus a third individual were still drifting around the island and had not yet settled into territories. Meanwhile, the moose population, which had jumped to more than 2,000 as wolf numbers had dwindled, was now showing signs of declining slightly. A consequence of the new wolf population? Too soon to tell.

My initial misgivings about intervening in natural affairs in a national park wilderness aside, the wolf translocation project couldn't be happening in a better place than Isle Royale. All the advantages that Durward Allen recognized on Isle Royale in the first place—its relative isolation from outside influence and its streamlined animal community—also make it an ideal outdoor laboratory for the reintroduction of wolves.

Plus we have six decades of data on wolves and moose there already, much of it collected by Rolf Peterson and John Vucetich, who continue to work on the island.

Most ecological studies are mere snapshots in time. At most they may last a season or two. When Dr. Allen set up the Isle Royale project, he imagined it might last ten years. But when ten years had passed, he realized the value in continuing to watch, to ask questions, and to learn.

As a result, his students and successors have seen and learned things they might never have imagined or witnessed had they just packed up their equipment and headed back to their offices for good. They might have left with the erroneous belief that they had learned enough. They might have left with the arrogance that they knew how things worked, rather than the humility that comes from realizing that nature is always capable of pulling off something unexpected. Rather than being confident that we can predict nature and thus control it, we are left with an appreciation of how little we can predict and how much we have to learn.

It is a sentiment that I'm sure Dr. Allen shared and would have appreciated.

At sixty-two years and counting, the Isle Royale project is the longest continuous study of any predator–prey system in the world. Even that is not long enough, and I'm sure that with more time of watchful waiting we will continue to learn more about wolves and moose in particular and the intricacies of nature in general. There is a rock-solid team of researchers at work. Funding seems secure, at least for the time being. We are poised to continue the study far into the future.

To me it doesn't really seem that long ago when it all began, and I took my first steps as a budding wolf biologist. That was June 30, 1958, my first day on Isle Royale. I hiked 7 miles from Rock Harbor to the Daisy Farm campground, and along the trail, my field notes recorded, I found "a fresh wolf track and 2 old droppings." Now, six decades and untold miles, wolf tracks, and droppings later, I am a mature wolf biologist. Instead of hiking 7 miles to catch a glimpse of wolf sign, I can look at a smartphone and check a GPS trail of a wolf online. I have changed a bit, research technology has changed quite a bit, and even the climate has changed.

What hasn't changed is that far out there in majestic Lake Superior still lies Isle Royale, a unique 210-square-mile wilderness laboratory, where wolves and moose await another youth eager to find his or her first wolf track and begin a lifetime adventure.

ACKNOWLEDGMENTS

So many people deserve thanks for their help and companionship during my time on Isle Royale some sixty years ago that it is difficult to remember and thank them all. Most, of course, have passed away. Nevertheless, I want to acknowledge those who were especially important and still stand out. At the top of the list is Purdue University's Dr. Durward L. Allen, who conceived the project and recruited me to start it at the behest of Dr. Oliver H. Hewitt, my undergraduate advisor at Cornell University. Next would be Dr. Robert M. Linn of Isle Royale National Park, who significantly promoted the project and loaned me his boat for the duration, and Donald E. Murray, who safely piloted me during almost all of my air time and was so critical to the success of the project.

Many other folks need to be recognized. Everyone from the Isle Royale rangers Ben Zerbey, Roy Stamey, Dave Stimson, and Pete Parry, as well as workmen Frank Taddeucci, John Murn, and Dave Kangas, to friends Pete and Laura Edisen, Phil Shelton, and Bob Janke: all contributed to the research and were great company.

Special mention goes to my ex-wife and still friend, Betty Ann Addison, who tolerated living in "the chicken coop," foot-pedal-starting the old Maytag washing machine, and keeping the kerosene lamp burning in the window in constant hope that through the dark fog my boat would appear. Also to my first- and second-born offspring, Sharon and Stephen, who were too young to remember but selflessly fed innumerable mosquitoes and blackflies as they played as little tykes along the Isle Royale shore.

Although they could have played no role in the early days of the Isle Royale research, Drs. Rolf Peterson and John Vucetich have contributed so much to the long-term success of the wolf and moose investigation. I thank them for the dedication they and their spouses, Carolyn and Leah, have afforded this project. For an excellent and thorough view of the Isle Royale wolf and moose research, please visit their outstanding, detailed, and informative website, https://isleroyalewolf.org.

This account was written primarily by science writer Greg Breining, who interviewed me extensively and consulted my complete journals, field notes, data forms, and other books and articles that I and others have written about the subject. Greg researched Isle Royale and the study, and he added immeasurably to the background detail as well as greatly enhancing the presentation, while I checked and approved the material, added a few anecdotes, and lightly edited the text as we went along. In a few places we included dialogue, which was not recorded contemporaneously but represents my best recollection of what was said under circumstances I did record in my notes. We thank Dr. Rolf O. Peterson and Dr. Michael E. Nelson for helpful suggestions for improving the manuscript.

When I began the Isle Royale work, little did I know that my first estimate of wolf numbers on Isle Royale would be followed by sixty others (as of this writing) or that the ninety-seven moose tooth rows and 438 wolf scats I gathered would start a collection now numbering in the thousands as later researchers added to my work. What I did know was that, for me, this was a chance of a lifetime, and I was going to make the most of the opportunity. I hope that this book shows that I tried my best.

INDEX

Index

otters, 12, 75, 118, 125, 126
Our Wildlife Legacy (Allen), 4, 10, 161
Outdoor Life, 114

parasites, 124, 136, 150, 162, 173
Passage Island, 98
Pennsylvania Game News, 22
peregrine falcons, 125
pesticides, 125
Peterson, Carolyn, 160
Peterson, Rolf, ix–xi, 5, 159–60, 165, 170, 171, 174; on ecological science, 163; longevity/continuity and, 160; parvo and, 163
Pigeon Point, 39
Pimlott, Douglas, 15, 16, 33, 123
Point Houghton, 52
politics, 151; life philosophy and, 152–53; religion and, 152
population dynamics, 5–6, 96
Port Arthur, 28, 40
predator-prey systems, 109, 161, 165; studying, 164, 175
predators, 63, 125, 141, 155, 169; population of, 165
prey, 5, 61, 63, 155, 163; alternative, 164; population of, 165; primary, 4; pursuing, 121
Purdue University, 10, 21, 22, 37, 41, 65, 69, 82, 111, 145, 151, 152, 153, 159; graduate work at, 4, 7, 13, 14, 19

rabbits, 8, 14; cottontail, 150, 153
radio collars, 4, 157, 160, 163, 164, 173
Rainbow Cove, 143, 144
Rainbow Point, 60, 101
Ranger II (boat), 15, 19, 23, 37
Ranger III (boat), 67, 82, 111, 127, 149; windy weather and, 112

ranger stations, 23, 40, 41, 43, 50, 51, 68, 79, 94, 129, 162
ravens, 31, 45, 48, 122, 137; scavenging by, 165; wolves and, 142–43
Recovery Plan for the Eastern Timber Wolf (Mech et al.), 166
red foxes, 31, 43; arrival of, 105; hunting by, 105; wolves and, 143
red squirrels, 31, 42
religion, 28, 180, 111–12, 151–52; politics and, 152
research program, 144, 153, 157, 161; beginning of, 5, 7–8; starting, 13–14
Richards, Paul, 79
Roberson, Marvin, 171
Robinette, Les, 147–48
Robinson Bay, 50, 143
Robinson Crusoe (Defoe), 25
Rock Harbor, 11, 19, 27, 37, 50, 54, 55, 79, 81, 112, 116, 145, 175; described, 23; fishing at, 114–15; moose at, 120; Sunday mass at, 28, 80, 151; wolves at, 98
Rock Harbor Lodge, 19, 23, 45, 70, 71, 81, 113, 146; church at, 151; meeting at, 152
Rock of Ages, 27
Rose, Bob, 15
Ross Island, 79
Rude, Elaine, 78, 80, 118
Rude, Mark, 80
Rude, Sam, 50, 51, 78, 80, 118, 119
ruffed grouse, 31, 76

St. Ignatius Loyola Catholic Church, 111
St. Paul Pioneer Press, 47
Sargent Lake, 23
scats: analyzing, 33, 73, 95, 174;

185

Index

L. David Mech is a senior research scientist with the U.S. Geological Survey and an adjunct professor in the Department of Fisheries, Wildlife, and Conservation Biology and Department of Ecology, Evolution, and Behavior at the University of Minnesota. Since the publication in 1970 of his book *The Wolf: The Ecology and Behavior of an Endangered Species* (Minnesota edition, 1981), he has been recognized as one of the world's leading authorities on the gray wolf. Among his many other books are the coauthored *The Wolves of Denali* (Minnesota, 2003); *Wolves: Behavior, Ecology, and Conservation*; and *Wolves on the Hunt*. He is founder of the International Wolf Center in Ely, Minnesota, and a member of its board of directors.

Greg Breining has written about science, nature, medicine, and travel for the *New York Times, Sports Illustrated, National Geographic Traveler, Audubon*, and many other publications. He is the author of more than a dozen books about science and the outdoors, including *Wild Shore: Exploring Lake Superior by Kayak* (Minnesota, 2000) and *Paddle North: Canoeing the Boundary Waters–Quetico*. He has taught at the Loft Literary Center in Minneapolis.

Rolf O. Peterson is a research professor in the College of Forest Resources and Environmental Science at Michigan Technological University and author of *Wolf Ecology and Prey Relationships on Isle Royale* and *The Wolves of Isle Royale: A Broken Balance*.